Dialogues on the Human Ape

CARY WOLFE, SERIES EDITOR

48 Dialogues on the Human Ape
Laurent Dubreuil and Sue Savage-Rumbaugh

47 Elements of a Philosophy of Technology: On the Evolutionary History of Culture
Ernst Kapp

46 Biology in the Grid: Graphic Design and the Envisioning of Life
Phillip Thurtle

45 Neurotechnology and the End of Finitude
Michael Haworth

44 Life: A Modern Invention
Davide Tarizzo

43 Bioaesthetics: Making Sense of Life in Science and the Arts
Carsten Strathausen

42 Creaturely Love: How Desire Makes Us More and Less Than Human
Dominic Pettman

41 Matters of Care: Speculative Ethics in More Than Human Worlds
Maria Puig de la Bellacasa

40 Of Sheep, Oranges, and Yeast: A Multispecies Impression
Julian Yates

39 Fuel: A Speculative Dictionary
Karen Pinkus

38 What Would Animals Say If We Asked the Right Questions?
Vinciane Despret

37 Manifestly Haraway
Donna J. Haraway

36 Neofinalism
Raymond Ruyer

35 Inanimation: Theories of Inorganic Life
David Wills

34 All Thoughts Are Equal: Laruelle and Nonhuman Philosophy
John Ó Maoilearca

33 Necromedia
Marcel O'Gorman

32 The Intellective Space: Thinking beyond Cognition
Laurent Dubreuil

31 Laruelle: Against the Digital
Alexander R. Galloway

30 The Universe of Things: On Speculative Realism
Steven Shaviro

29 Neocybernetics and Narrative
Bruce Clarke

28 Cinders
Jacques Derrida

27 Hyperobjects: Philosophy and Ecology after the End of the World
Timothy Morton

26 Humanesis: Sound and Technological Posthumanism
David Cecchetto

25 Artist Animal
Steve Baker

(continued on page 232)

Dialogues on the Human Ape

Laurent Dubreuil and Sue Savage-Rumbaugh

posthumanities 48

University of Minnesota Press
Minneapolis
London

Published by the University of Minnesota Press
111 Third Avenue South, Suite 290
Minneapolis, MN 55401-2520
http://www.upress.umn.edu

Printed on acid-free paper

Library of Congress Cataloging-in-Publication Data
Names: Dubreuil, Laurent, author. | Savage-Rumbaugh, Sue, author.
Title: Dialogues on the human ape / Laurent Dubreuil and Sue Savage-Rumbaugh.
Description: Minneapolis : University of Minnesota Press, [2019] |
Series: Posthumanities ; 48 | Includes bibliographical references and index.
Identifiers: LCCN 2018024169 (print) | ISBN 978-1-5179-0564-4 (hc) |
ISBN 978-1-5179-0565-1 (pb)
Subjects: LCSH: Apes—Psychology. | Animal communication.
Classification: LCC QL737.P9 D69 2019 (print) | DDC 599.88—dc23
LC record available at https://lccn.loc.gov/2018024169

UMP LSI

Contents

Authors' Note

The raw material for this book consisted of actual dialogues that took place in 2014 and 2015. The recorded exchanges were later transcribed by Joseph Fridman, with the help of Faye Elgart. They were shaped into book form by Laurent Dubreuil in 2015 and 2016. The final text was revised and expanded by Laurent Dubreuil and Sue Savage-Rumbaugh from 2016 to 2018. The narrative voice in the opening and closing of each chapter is that of Laurent Dubreuil; he also wrote the Introduction, featuring quotations from written material prepared by Sue Savage-Rumbaugh. A synthetic "Timeline of Ape Language Research," coauthored by Savage-Rumbaugh and Dubreuil, appears as an Appendix at the end of the book.

Endnotes provide direct references as well as further explanations or comments, though the latter have been deliberately kept to a minimum. These notes are signed with the initials of the author, SSR or LD.

Acknowledgments

The Mellon Foundation provided important financial support for the authors' travels and for the transcription of their recorded discussions. A special grant from the Region of Catalonia also helped defray the travel and lodging expenses of Sue Savage-Rumbaugh to Ithaca in November 2014. Ryan Sheldon graciously hosted the two authors in August 2014 in Osceola, Missouri.

For their helpful comments on various states of the manuscript, we heartily thank Douglas Armato, Cathy Caruth, Laurent Ferri, Joseph Fridman, Russ Tuttle, Cary Wolfe, and an anonymous reviewer. Laurent Dubreuil is grateful to all the persons who made his past visits to the lab possible, and especially Liz Rubert-Pugh, who, over the span of three decades, played an absolutely essential role in the experiment and in the physical, emotional, and intellectual well-being of the bonobos. Sue Savage-Rumbaugh wishes to thank all former and current members of Bonobo Hope and of the legal team, in particular Sally Coxe, Julie Flaherty, David Goldman, Ursula Goodenough, Nancy Howell, Aya Katz, Sheila Knoploh-Odole, Carmen Maté, Ross Neihaus, Douglas Pernikoff, Itai Roffman, Liz Rubert-Pugh, Duane Rumbaugh, Shane Savage-Rumbaugh, Ryan Sheldon, Paul Thibault, Derek Wildman, and William Zifchak.

Of course, this book would not exist without the bonobos. Sue is dedicating *Dialogues on the Human Ape* to the memory of the bonobos Panbanisha, P-suke, Nathan, Tamuli, and Matata, and the chimpanzees Lana, Sherman, Austin, and Panzee—who all gave their lives with exceptional wit, vigor, and great enthusiasm to the understanding of language, the *Pan/Homo* dynamic, and the way we are. Sue writes: "They all left too soon, as did Duane

Rumbaugh, who guided this work from its inception with Lana the chimpanzee in 1970 till his death in 2017. He is survived by Mercury, Kanzi, Nyota, Teco, Elikya, and Maisha—all of whom desire to continue their contributions while they remain here, and all of whom recall all symbols they have learned and seek to use them, even when actively discouraged and/or prevented from doing so. Such is the mark of a 'truly languaged being.' "

Introduction

While the desire to engage in conversations with other animals is potentially a feature of most, if not all, human civilizations, the more particular quest for teaching our language to great apes is tied to the rise of science in modern times. As soon as living apes are introduced in Europe, the idea of using systematic methods to allow verbal communication with those specimens is being brought forward. Advances in teaching deaf children were already conceived during the Enlightenment period as a model to emulate with other primates. The first serious attempts at granting human language to nonhuman apes date back to the early twentieth century; at the same time, some young chimpanzees survive long enough out of Africa to learn to do music-hall routines, thereby influencing writers such as Franz Kafka (for his "Report to an Academy") or movie directors such as Buster Keaton (for his *Playhouse*). From the 1910s to now, individuals from all four nonhuman species of great apes (orangutans, gorillas, chimpanzees, bonobos) have been put to the test.

Different interfaces for language production in apes have been tried: *direct, oral, speech* (which comes with limitations owing to anatomical constraints on the vocal tract in nonhuman primates); *sign language*; *colored—and supposedly semantic—tokens*; arbitrary signs called *lexigrams,* representing words or grammatical elements on a computer keyboard (with or without a synthetic voice uttering the term corresponding to the key pressed by the ape); *writing and spelling.* The major breakthrough with the female chimpanzee Washoe in the late 1960s and early 1970s experiments first conducted by the Gardners and then by Roger Fouts happens through the use of sign language. This option is soon employed by Penny Patterson with the female gorilla Koko and by Lyn Miles with the male orangutan Chantek. Sign language, however, requires a knowledge of the particular ape dialect of American Sign Language (ASL) (as the anatomy of nonhuman primates does not allow for the exact replication of the standard sign), which often means that a level of speech interpretation has to be sought for, a feature that is open to critique. To avoid some of these difficulties, Duane Rumbaugh, Ernst von Glasersfeld, and their collaborators at the Yerkes Center in Atlanta implement an artificial language ("Yerkish"), using lexigrams and a specific syntax, on a computer keyboard first used by the female chimpanzee Lana in the early 1970s. Still, the whole field of "ape language study" is shattered in the late 1970s, when Herbert Terrace, in his own work at Columbia University with the male chimpanzee Nim Chimpsky relying on ASL, fails to replicate the results of other scientists. Speech production in apes using sign language is discarded as mere repetition of the previous gestures made by their human interlocutors, interspersed with "wild cards," randomly selected. The effects of Terrace's failed experiment are very powerful in the United States (where virtually all of this line of research has been pursued) and, as of the early 1980s, they contribute to making such endeavors very difficult to fund.

In 1970, Sue Savage-Rumbaugh, a recent college graduate who has been admitted to Harvard and is on her way to study psychology with behaviorist B. F. Skinner, happens to stay with a friend in Norman, Oklahoma. She meets Roger Fouts at the University of Oklahoma, who invites her to the "Chimp Farm." Spending three full days with the apes, Savage-Rumbaugh decides to reject the offer from Harvard, as Skinner has overlooked chimpanzees and the role of language in their development when he formulated and refined his theories of human behavior. During her graduate years in Oklahoma, Sue works with chimpanzees having very different rearing backgrounds and life experiences (some individuals being hosted in human homes and treated as family members, others living in groups of conspecifics on an island, etc.), and she is exposed to a wide array of behaviors in diverse primate and animal species. She teaches sign language to many chimpanzees. In 1975, after receiving her doctoral degree, she moves to Atlanta to work as a postdoctoral fellow in the laboratory of Duane Rumbaugh, whom she later marries. In the early 1980s, Savage-Rumbaugh, now on the faculty of Georgia State University in Atlanta, works at the recently founded Language Research Center (LRC). She uses computers and lexigrams in her experiments with the two male chimpanzees Sherman and Austin. A new, spectacular, breakthrough in ape language comes in 1985 when Kanzi, a young male bonobo, spontaneously acquires an impressive command of lexigrams by attending the training sessions held by Savage-Rumbaugh and collaborators for his adoptive mother Matata. Building on a new paradigm made of immersion within symbolic culture and a constant use of language through both scripted and improvised life events, Sue pursues the trajectory with Kanzi and integrates new participants, such as a female bonobo named Panbanisha (Matata's daughter) and a female chimpanzee named Panzee. In 1993, Savage-Rumbaugh shows that Kanzi's understanding of spoken English is on a par with that of a human child. In the late 1990s, both Kanzi and Panbanisha display a remarkable comprehension of the lexigrams and of spoken English. They also vocalize specific sounds corresponding to English words (such as *right now*) and they sometimes draw a lexigram instead of pointing at it. While those latter abilities are manifest in daily life and well documented in the video archive, they are not systematically explored in publications, as other aspects are concurrently investigated, such as the fabrication and use of tools, counting capacities, the existence of a "theory of mind," temporal processing, fire making, self-representation, and the like. Epistemologically speaking, Savage-Rumbaugh moves toward an interdisciplinary blend of anthropology and experimental psychology that she calls *experimental anthropology*. In her collaboration during the late 1990s and early 2000s with interlocutors such as William Fields, who helps her in raising Nyota (Panbanisha's son), Sue focuses on the role of culture as the driving force behind the behavioral changes and cognitive extensions that take place in the apes she works with.

Six lexigrams. From left to right: same, with, surprise, Sue, ? (question), put.

In the first years of the twenty-first century, Georgia State University sells the bonobo colony to Ted Townsend, a philanthropist from the Midwest. In 2005, all the bonobos—the "languaged" individuals as well as the members of the control group—are relocated to Des Moines, Iowa, where the Rumbaughs and Sue's sister Liz Rubert-Pugh are hired by Townsend to continue their work in a new, vast, private laboratory. The promise is to maintain the research and the ape group for life. The move, however, is not felicitous, as other, nonacademic and financial, interests come into conflict with the objectives of scientific research. The Iowa Primate Learning Sanctuary (IPLS), doing business as the Great Ape Trust, quickly becomes an operation where Savage-Rumbaugh's input is marginalized at best, negated at worst. In particular, zoologically motivated interests become tightly linked to the founder's desire to build a large public attraction. Sue's decision to list Kanzi, Panbanisha, and Nyota as coauthors of a 2007 peer-reviewed article where she asks the apes their opinion about their living conditions in captivity causes a stir in her own institution (an "affair" we return to at the end of our last dialogue). Sue's standing in the lab is fragile, and she is fired, then reinstated, though as a participant with "special standing" but no direct authority. In 2008, a major flooding of the facility threatens the lives of the bonobos, who are stranded in the laboratory for days, with Sue Savage-Rumbaugh and Liz Rubert-Pugh spending most of their time with the apes and making the liaison with the outside world. In 2010, the birth of the male bonobo Teco gives Savage-Rumbaugh the opportunity to come back to the laboratory on a daily basis, as the staff (except for Liz) has no expertise in the matter. Sue is implementing a new rearing protocol with Teco.

In September 2010, I meet Sue Savage-Rumbaugh for the first time in Iowa to spend a week at the lab. A philosopher and literary scholar by training, I have always been drawn to interdisciplinarity and I have written extensively on the powers and limits of language and thought. Thanks to a "New Directions" fellowship awarded to me by the Mellon Foundation, I am spending the whole of 2010 studying how experimental psychology, cognitive science, artificial intelligence, paleoanthropology, and formal logic approach "language." This line of research in my work is later represented by my 2015 book *The Intellective Space,* an "undisciplined" essay exploring how the very performance of cognition is apt to bypass the boundaries of thought. Besides intellectual curiosity, the impetus of my 2010 visit to the Trust is twofold: to get a—deliberately subjective—hold on the kind of verbal exchanges humans could have with the "languaged" apes (thus complementing my more theoretical study of the research on primate cognition and communication), and to discuss more with Savage-Rumbaugh the singularities of her own method. (Incidentally, in 2000, as I was a graduate student in literature and philosophy, I had already begun to read some of Sue's work, after watching on French television parts of the NHK documentaries on Kanzi.)

At the end of my first week at the Great Ape Trust, it becomes apparent to me that Kanzi, Panbanisha, and Nyota are partaking in human language on a more sophisticated level than I had once thought possible, and that the experiment is now going much further than "just" ensuring verbal communication between human and nonhuman apes. Several conversations with Savage-Rumbaugh also confirm an impression I had in reading her more recent writings: She has largely been constrained by the standardized protocols of scientific publishing, her own research and elaboration having somehow outgrown the set limits of her field. Both in 2010 and in 2011 (my second, and longer, visit to the lab in Iowa), Savage-Rumbaugh and I establish a habit of dialoguing about language and what it allows the speaker's mind to achieve—those discussions relying on multiple viewpoints, references to daily human activities, ape experiments in a lab, or Greek philosophy and literature. These informal dialogues, often several hours long, continue to take place over the years, with occasional contributions by the bonobos if our exchanges take place in the lab.

In November 2011, I send an electronic message to Savage-Rumbaugh about a book project we could write together. It is tentatively titled *Dialogues between Apes: Language, Life, Science, Fiction.* Here is what I suggest: "This book is about dialogue, i.e. about the transmission of language and reason *(dia logon)* among eloquent apes. In accordance with this idea of dialogue, each contributor retains her or his own authorship on the texts she or he writes, which allows for a greater plurality of viewpoints. The volume is conceived as an experimental work, written in collaboration by humans, with the help of bonobos. The horizon is a fresh take on the co-creation of truth and fiction within 'humanness,' and the methodological project generally lies at the crossroads of the Humanities and the Sciences."[1] Sue is enthusiastic. Unfortunately, the project needs to be put on hold for several years.

As announced a few months earlier, Ted Townsend stops funding the laboratory in 2012. The loss of such a vast financial support immediately puts tremendous pressure on all participants, as the facility is very expensive to run. Financial downsizing also means laying off many employees, cutting the salary of others and relying more on volunteers (including, soon, Savage-Rumbaugh herself, who will spend her personal savings to cover some of the costs at the lab). Later in 2012, a dozen former employees accuse Savage-Rumbaugh of misconduct with the apes and state that she is mentally unfit. As the allegations are under review, she is put on administrative leave but permitted to remain on site as long as the veterinarian is present. Panbanisha passes away in November of the same year. After the death of Panbanisha and after the accusations of misconduct are dismissed by the review committee, Sue is reinstated. I join the board of directors of IPLS at the end of 2012.[2] Through a settlement agreement executed in February 2013, all bonobos except one become co-owned by the Iowa Primate Learning Sanctuary and Bonobo Hope Initiative (BHI), an organization created by Sue Savage-Rumbaugh. IPLS is tasked with raising funds for the operations in the facility it owns, while Bonobo Hope oversees with Sue the research trajectory and the bonobos' welfare. With the exception of Sue, the academics formerly associated with IPLS (myself included) are being advised to step down and move to the board of BHI.

In 2013, Sue falls and suffers a concussion that keeps her out of the laboratory for five months. In the summer and autumn, BHI discusses ways of seconding her in her work and finally decides to appoint two of her former students at Georgia State University (Jared Taglialatela and William Hopkins). Right after the vote, Taglialatela and Hopkins seize control of the laboratory. Whereas in May 2013 the board of IPLS had resolved to guarantee Savage-Rumbaugh unfettered access to the bonobos for life, as of November Sue is banned indefinitely from the facility and can no longer see the bonobos or even exchange with them over video from a remote location. The two new directors of the organization replacing IPLS (now the "Ape Cognition and Conservation Initiative" or ACCI) claim that Sue and Bonobo Hope have relinquished all authority. Despite co-ownership, BHI scientists are only allowed in the facility for very brief and infrequent visits. During a half-day visit I could make in 2014, I see, as others have witnessed before and after me, that the bonobos are no longer *immersed* in the symbolic world they used to live in, and that linguistic interactions are now severely limited, both in scope and in quantity. Caretakers are constantly kept separate from the apes by a wire and they wear gloves and masks, impeding nonverbal communication through facial expression. This is what Sue calls "Yerkes lite," by allusion to the working protocols with great apes employed in medical research at the Yerkes Center.

In 2015, and after more than a year of unsuccessful attempts on the part of Bonobo Hope to reach a workable agreement with the other side, the only way to resolve the matter appears to be a federal trial to be held in May in Des Moines. Bonobo Hope and Sue are asking for the restoration of the research trajectory, legally defined in the side agreement as "research in the fields of experimental psychology; use of language and tools; and ape intelligence and human cultural modes (including but not limited to art, music, tools, agriculture, fire, animal domestication, habitat construction, use of water, hunting, mimits *[sic]*, sociological role construction, normative child rearing practices, play or normative religious/mythological practices)." At the hearings, I, along with other academics, testify in favor of Sue, BHI, and the bonobos. A project for relocation of the bonobos to a low-key but more adapted facility, to be financed by the entrepreneur Ryan Sheldon (one of Sue's former collaborators at the LRC), is presented in court. In November 2015, the judge declares that he lacks jurisdiction and that he cannot adjudicate the claims of Bonobo Hope, later prompting BHI, in 2018, after another failed round of negotiations with ACCI, to move to state court in order to correct the situation and help the five bonobos (Kanzi, Nyota, Teco, Elikya, Maisha) who still reside in Iowa. In the midst of all these events, and beginning in the summer of 2014, both Sue and I manage to find time to work on the *Dialogues.*

Aside from such factual considerations, this book is first and foremost an intellectual endeavor. Our *Dialogues* deals with the theoretical and practical dimensions of what being a "human animal" means. This requires a few preliminary suppositions. First, we admit that, despite the vagueness of definitions, there is something human, and that this "something" is being *affixed* to animal beings (our ancestors, ourselves, nonhuman primates such as Kanzi). Humanness is a process, at both the individual and the group levels. The epigenetic (deriving from a combination of culture, individual development, and ecological circumstances) makes *Homo sapiens*—and any other animal—more than the sum of its genes. The human could be located as a "self-specification," governed by the discursive and the imaginary, having effects on organisms across generations. Thus, humanness is neither the automatic consequence of a genetic profile nor a mere construction of "free-floating" (and "unnatural") cultures. It is a *possibility* that cannot arise in any species; but, as a possibility, it is able to emerge in a priori "nonhumans" (bonobos, or *Pan paniscus,* and chimpanzees, or *Pan troglodytes,* being among the most obvious candidates at the species level).

The debate about the "human animal" is being obscured by many factors. A profound resistance to the fact that we are not "above" animality, and, conversely, a subsumption of all differences across species (and individuals) into the rather mythical category of "the animal" (or of "sentient beings" in general) are metaphysical obstacles to our inquiry. Positing that humans are animals as far as their basic needs of subsistence and reproduction are concerned, but not for their higher-order activities, is simply wanting. If our genes shape and limit what "we" do (and are able to do), it is no longer possible to discard the effects of individual and social development or the way the epigenetic is able to override genetic "programming." But an additional difficulty derives from the compartmentalization of knowledge. While contemporary primatology and psychology have some obviously philosophical dimension, they are often underconceptualized or dogmatic. Similarly, many scholars in the "posthumanistic" field of "animal studies" tend to overlook what nonhuman animals are apt to tell us about themselves (and about us). There is a crucial need, then, for a meaningful dialogue across the disciplines that would be more than an honest conversation or an encyclopedic assemblage of viewpoints.

As a philosophical genre, *dialogue* is closely tied to the reconfiguration of philosophy after Socrates. Far from being a now-outdated way of spreading ideas and theories, dialogue is, I believe, a powerful tool for the dissemination of speculative hypotheses and for open-ended debates. In its different forms, dialogue is even a structure of psychogenesis. If there is not "one" definitive truth about the nature and possibilities of the human animal, a dialogic description is certainly "truer" than one-sided statements about the alleged inabilities of nonhuman primates. In accordance with the Platonic tradition, I also weave into our *Dialogues* moments of description and narration—because abstracting scholarly thinking from the very facts of life is not an ultimate virtue. And because this book is polyphonic, I am now letting Sue add, in her own voice, several important considerations about the rearing method she developed with the apes, her approach to language "in action," and the theoretical conclusions she reached about humanness. She writes:

> We become a part of what is occurring by becoming a part of a dialogue. A dialogue between Sue and Laurent allows the reader to identify with one or the other and thus to become a part of the language action, so to speak. Dialogue also allows for natural digression and relating back to the world outside the dialogue, thus grounding it in a place, a time, and a situation. Lastly, *the language of*

Kanzi and family was brought into existence by the process of dialogue embedded in real-life contexts and events—not by planned experimental presentation. The reason Kanzi began to speak—all on his own—was to bring coordination to thought and action in a way that he desired in his world—this is something only dialogue can do. At its most basic level, language is contextually embedded dialogue as it first emerges. It cannot, does not, and will not arise or manifest, initially, in any other manner. The first exchanges are always pregnant with "meaning making" in the midst of contextually embedded dialogue. As meaning is made, two or more participants begin to coordinate their thoughts, views, perspectives, goals, and perceptions. As this happens, the child (or ape) becomes a participant in the dance of common cojoined awareness and takes the initiative to co-structure what is going to take place, through language.

For example, Kanzi loved to be chased. He initiated this by a smile, touching a person, and running away while looking back. The person responded by saying "YES CHASE KANZI," and did so.[3] Kanzi then began to say "CHASE" and look at the person with a questioning expression. But Kanzi did more. He wanted person X to chase person Y. So he began to point to A, then touch "CHASE," then to B. In so doing he abstracted himself from the chase. And "CHASE" itself became an action that connected other participants (which might or might not be Kanzi). This happened very early, by the time Kanzi was 2.5. This "idea" was Kanzi's, and the linguistic arrangement was Kanzi's—no person had this thought or utterance. Kanzi did this because he wanted to see the social relationships between human beings become manifest through a game of chase. This was something he could understand. The subtle social hierarchy between humans was not otherwise visible to him. And people were often staid and noninteractive. This got them moving and interacting and expressions began to emerge on their faces. Kanzi found this to be lovely.

There are no reports of other "languaged" apes such as Nim, Washoe, Lucy, Koko, or Chantek abstracting their language use in such a way. Essentially all descriptions of their utterances fall into two categories: a) answers to questions directed to them, b) requests they produce for goods and services directed to them. This limitation in their language arose from the limitation of their dialogues, and the dialogue's limitations arose from the limited expectancies of the human participants. Their dialogues always involved questions or requests. They used these kinds of dialogues to "get the apes to sign" as these apes were often silent. Nim, Washoe, Chantek, Lucy, and Koko did not have a forest world to traverse and explore, as Kanzi did. The constant emphasis on teaching and "what this?," "what that?"—made them hesitant to use language for its true purpose: telling someone something you want them to know and that you have reason to believe they don't know. This purpose arises in the actual living of life; it has no raison d'être in training studies. The Gardners understood this and tried to create, as best they knew how, a "real life" for Washoe. But that "real life" became focused around "collecting data on Washoe's utterances" and creating situations to get Washoe to sign. This was what graduate students were to turn in at the end of the day for analysis.

Lacking the kind of life and dialogue that Kanzi experienced, other apes failed to utilize language to abstract away from the self. By 2.5 Kanzi was more interested in others than in the self and he was dialoguing with them in a way that enabled him to transition to seeing and employing language as a vehicle for "aboutness." This interest in, and focus on, "the other" is quite natural for apes, but they must be reared in social group for it to proceed normally. Kanzi was lucky—he had mother Matata. The other language-using apes had no ape mother. Of course, Kanzi had his mother by design and intent. I made sure of that. Before I agreed to present him with language, I worked with Matata. No other ape language researcher ever tried to teach language to an adult ape. By the time Kanzi was born, Matata and I had been good friends for over 5 years. And bonobo mothers, unlike gorillas, orangutans, and chimpanzees, share the burden of infant care. So Matata was happy for Sue's help and for Liz's help. We never "took" Kanzi from Matata as some have claimed. Kanzi jumped into our arms whenever he wanted and he jumped back to Matata whenever he wanted. Thus began the bicultural rearing of Kanzi. It took place in 50 acres of forest, with Matata and Kanzi's life was exceptionally different from that of any other ape who acquired language. This mattered.

A social group can be a mixed-species group, as was the case with Kanzi, Panbanisha, Nyota, Nathan, and Teco. Rearing an ape with humans alone is not sufficient, because we are unable to process sounds, images, and social behavior at the proper rate. For the ape nervous system to develop normally, interaction with other apes is required. Similarly, if humans were reared only by apes, many of our capacities would not develop normally. This mixed-species world was not based on control of Matata or Kanzi. It was based on cooperation and love. The humans did not punish apes, but Matata sometimes punished humans when they did not abide by her bonobo rules.

Most ape language efforts after Kanzi stumbled toward the awareness of the value of having other chimpanzees present. This immediately facilitated linguistic capacity, but only with Kanzi (and the bonobos who followed him) was this process intentionally and deliberately employed within the nexus of a "created life." Kanzi's life was not instrumental or planned. It was as spontaneously created each day, in a novel and meaningful and ordinary way—as would be the life of any human child. Like all human life—it was a symbolically created reality. And within that reality bubble Kanzi and the other bonobos found a place. But it was not to last; the relocation to Iowa ended the access to forest, though this was never the intent. As that access ended, the feeling of being captive set in.

> It was this life and the process of dialogue that drove language acquisition and use in the bonobos. Kanzi spoke because there were real things to speak about and people to speak with (the dogs that appeared in the forest, the rain, the fire in the play-yard, the visitors, etc.). It was this process that rendered the bonobo's language capacities so like our own. *And it was their internal development of language that took off near adolescence and began to create bicultural beings with their own thoughts and perspectives on life.* Their world, the move itself to Iowa, the "new" people that "worked with them," and the fact of captivity itself all acted together to change them in fundamentally deep ways. And they began to learn from TV about the killing of apes in Africa and to reflect on this. For Nyota this was very difficult. He liked books and he wanted books all around all the time—but this was not "permitted." In Iowa, they learned that most humans did not think apes capable of "real" language, and for Panbanisha, movies like *Harry and the Hendersons* exemplified her predicament. It was—to make the understatement of the century—astonishing to watch what language enabled them to do and think as maturity took hold. They were becoming truly *human,* exquisitely *human,* morally *human* and philosophically *human.*[4]

This would not be my own perspective exactly, as, for instance, I don't know what being "truly human" could mean. Or I would be more careful with distinguishing the varied layers of dialogue, because inner differences in the disposition of verbal exchange exert vast consequences over the way we think. But, precisely, it is because I have always been invested in contradiction, becoming, and signification—both as a scholar and as a "person"—that I enjoy the movements of convergence and divergence, of unison and discordance that are consubstantial to dialogue (in its maximal sense). Getting the dialogic conditions right is, in itself, a demanding task. Our current political, social, technical environment is not particularly propitious and favors identity-based soliloquies. Then, in any actual exchange, our sensitive selves are necessarily *touched*: we all carry with us the imprint of our past affective experiences, and we are sometimes unable to productively channel our emotional life into our thinking as it relates to other, embodied, noetic agents. In fact, there is always some miracle when an intense, creative dialogue takes place. In our case, Sue attributes a lot of the "magic" to the felicitous interactions I had with the bonobos, and to their reactions to me. In a sense, she argues that they showed, and told, her she could exchange with me. This a matter *I* cannot really speak about, but she insisted our readers know more about this. She writes:

> It is the bonobos' acceptance of Laurent that enabled him to begin to understand their world. And this acceptance took place through and because of the type of dialogue and "presence" that Laurent extended to the bonobos themselves. The standard view—that humans can never understand the mind of another species—was collapsed with Laurent through dialogue. Dialogue is language in action and, understanding this at a deep level, Laurent jumped into action with his words. This enabled him—in his dialogues with me—and with the bonobos, to discuss the concepts that undergird the work from the role of participant observer. And thus we could proceed into territory not yet covered by past treatises on "ape language."
>
> These have been limited by the writer's need for explanation from "their own worldview." It is this ever-present "need" that has limited the field, not the inherent capacities of the bonobos themselves. Laurent was intellectually prepared to "jump into language" knowing enough about it to search for, but not to require constant "proof" of, "mindful communicative intent." His intense sensitivity was at once understood and appreciated by the bonobos and doors opened. It is these doors, which require dialogue to open, that other writers have yet to try to address. Many others have felt compelled to write books and articles to explain to others what it means to literally talk to these bonobos. They have tried to say it means that

> you must treat them as you would treat a person—every moment—and in so doing you open a door in your mind that was previously closed. Saying this is one thing. It is dialogue that actually does it. Testing them doesn't do it. Asking them questions doesn't do it. Talking to them doesn't do it. Waiting for them to prove that they have language doesn't do it.[5]

Obviously, then, our dialogues had to make room for the thinking (and communicating) apes Sue Savage-Rumbaugh has reared. The original intent was to invite the bonobos to intervene in our exchanges at some key moments. Unfortunately, for the reasons I have described, we could not achieve this goal. For now, the human–ape dialogue must remain between Sue and Laurent, and mainly *refer* to other "languaged" beings. But, in Sue's words, "our 'bonobo hope' is that the reader's knowledge may help change this absurd state of affairs and restore the community of dialogue between bonobos and human beings to them and to science."[6]

The first dialogue in this book is a "warm-up," where we consider the stratified constructions of the category of the "animal." This initial chapter is also a reference to contemporary aspects of the question, within the fields of "posthumanism" and "animal studies" and in the ape legal-personhood movement. We later move on to an exploration of concepts and problems that are in relation with the qualification and experience of humans and animals. The second and third dialogues interrogate the notion of consciousness as well as its different types (or "flavors"), introducing ideas on their dialogic and symbolic dimensions among "languaged" animals. This leads us to focus on human language in the fourth chapter: how to approach it both theoretically and experimentally, how chimpanzees and bonobos could acquire and use it. In society, a symbolic mind, immersed in language and endowed with self-reflexive consciousness, manifests a will and should have rights. Our last dialogue ties together the positions we previously discussed by inserting them in a more straightforwardly political frame. After debating the roles and functions of determinism, this final exchange evokes the kind of duties and rights one could identify in or demand for "eloquent apes."

Some Primates Mentioned in the Dialogues

Austin (1974–1996). A male chimpanzee involved in a language experiment based on lexigrams and led by Sue Savage-Rumbaugh.

Elikya (born 1997). The daughter of Matata. A female bonobo involved in a language experiment led by Sue Savage-Rumbaugh; a member of the control group.

Roger Fouts (born 1943). A male human and psychologist whose research involved sign language and apes such as Washoe. He trained Sue Savage-Rumbaugh during her graduate years.

Beatrix Gardner (1933–1995) and her husband **Allen Gardner** (born 1930). A human couple who conducted an experiment with Washoe involving sign language.

Ernst von Glasersfeld (1917–2010). A male human and a researcher in cybernetics and linguistics. Collaborating with Duane Rumbaugh, he designed the initial version of Yerkish, an artificial language Lana was the first chimpanzee to use.

Catherine Hayes (1921–2008). A female human who, for an experiment designed with her husband Keith, raised the female chimpanzee Viki as a human child and taught her, through speech-therapy methods, to utter a few English words, as documented in the 1951 book *The Ape in Our House.*

Kanzi (born 1980). Matata's adoptive son. A male bonobo involved in a language experiment based on both lexigrams and spoken English, led by Sue Savage-Rumbaugh.

Winthrop Kellogg (1898–1972). A male human and psychologist who, with his wife Luella, briefly raised in the 1930s the female chimpanzee Gua together with their human child Donald, using the same rearing method for both individuals.

Lana (1970–2016). A female chimpanzee involved in a language experiment using lexigrams and led by Duane Rumbaugh.

Matata (1970?–2014). A female bonobo, involved in a language experiment led by Sue Savage-Rumbaugh using lexigrams and spoken English.

Nim (1973–2000). A male chimpanzee, first raised by Savage-Rumbaugh, then involved in an experiment using sign language led by Herbert Terrace.

Nyota (born 1998). A male bonobo involved in a language experiment based on both lexigrams and spoken English led by Sue Savage-Rumbaugh.

Panbanisha (1985–2012). Matata's daughter and Elikya's older sister. A female bonobo involved in a language experiment led by Sue Savage-Rumbaugh using lexigrams and spoken English.

Panpanzee, also called **Panzee** (1985–2014). A female chimpanzee. In the first years of her life, she was involved in a language experiment led by Sue Savage-Rumbaugh and she was raised as a sister to Panbanisha.

David Premack (1925–2015). A male human whose extensive career as a psychologist included a language experiment conducted in the 1960s with the female chimpanzee Sarah and "language" tokens.

Duane Rumbaugh (1929–2017). A male human who, for decades, conducted and supervised research in psychology involving primates, including the language work with Lana in the 1970s. He was associated with the experimental work led by his wife Sue Savage-Rumbaugh until the 2000s.

Sherman (1973–2017). A male chimpanzee involved in a language experiment based on lexigrams and led by Sue Savage-Rumbaugh.

Teco (born 2010). A male bonobo, the son of Elikya and Nyota. In the first years of his life, he was involved in a language experiment led by Sue Savage-Rumbaugh and raised as the son of Kanzi.

Herbert Terrace (born 1936). A male human whose current fame mainly derives from his failed 1970s psychological experiment with Nim and sign language.

Michael Tomasello (born 1950). A male human and psychologist whose research focuses on primate cognition and communication.

Washoe (1965?–2007). A female chimpanzee involved in a language experiment based on sign language, led first by Allen and Beatrix Gardner, then by Roger Fouts.

DIALOGUES

On Animals and Apes

We are beginning with the following description, originally written (in French) in 2010.

"I am with a dozen of strangers in the lobby of a brutalist edifice. In front of us is the other part of the building: we see it through the glass. There lives an entire family. I am waiting for something staged like an apparition. Through books and videos, I have come to know the different actors that are about to enter the stage, even though I am still unfamiliar with the size and plan of the lab I am in. This morning, I took a taxicab from the downtown area of the Iowa state capital to go to this place that is officially, and ambiguously, named 'The Great Ape Trust.' As will happen in each of the following days, the driver is stunned when I explain that I, the French professor in an Ivy league School, am in the Midwest to see chimpanzees using keyboards to communicate in English with humans. The Trust is a facility in the West Des Moines suburb; it is surrounded by a barbwire fence, with a checkpoint operated by a private company, whose employees escort visitors up to the building entrance. The lab has many doors that codes and access badges open and close. The guests are now waiting for the 'baby shower' to begin: Kanzi's son Teco is barely two months old, and he is going to be introduced to the world.

"The architects have designed an observation room with sliding doors that is like a foreign body within the ape enclosure: a room where bonobos can enter and be seen more easily by the humans staying in the lobby, a place where they can use the huge touch screen that is at their disposal. On the other side, a staff member has pressed the lexigram VISITORS, a word that automatically resounds through the loudspeakers of the lobby. From where we stand, we have no clear view of the outside. The sliding doors and the oversized computer screen could lead one to believe that we are aboard a spaceship, or, at least, on the set of a science-fiction TV series. What is staged here is futurity. But the grandiose setting contrasts with the familiar simplicity of the characters that are beginning to appear: Sue the scientist holding Teco in her arms, Kanzi the bonobo, his half sister Panbanisha, her son Nyota. As they come into the observation room, this is no longer sci-fi, just present. The apes are not going to deliver a series of prophecies. They rather try to be understood, to ask for the items they need, to make a comment or a joke. After the games of CHASE and TICKLE that are required by Nyota, a conversation begins. It is full of loud calls and screams, of deictic gestures, of lexigrams, of interpretive remarks delivered by Sue. Nyota and Panbanisha ask several times to go outdoors, which is no longer possible, for the electric fence has not yet been repaired since the 2008 inundation ('BIG BIG WATER,' Sue says). Nyota wants his JUICE. Panbanisha would like to sip a cup of the 'STAFF OFFICE COFFEE.' Everybody would like to open the SURPRISE-S.[1] There is much energy and some benevolent chaos."[2]

As I was discovering the different members of this group in September 2010, I did not intend to become so close to them, in heart and spirit. But the significance of our first encounter struck me quite rapidly. A few friends were even a bit intrigued by my new obsession with apes. As early as November 2011, I proposed to Sue to write a book of dialogues. Massive difficulties at the lab in 2012 and 2013 prevented us from moving forward. In the summer of 2014, Sue had now been excluded from her own research and separated from the bonobos for months.[3] She and I reconsidered the idea of having our dialogues, with a somewhat different content and a new modus operandi.

In November 2014, Sue was scheduled to come to Cornell to deliver a keynote lecture at a posthumanities conference I would also participate in. We seized this opportunity to make progress on our project. Before this, we already had had many dialogues together, at the lab, in the Midwestern countryside, at a Chinese teahouse, in a Tex-Mex fast-food restaurant, or on our way to a box storage facility. Those discussions were just for ourselves.

New conversations would occur now, in the perspective of this book. While we were both in Ithaca, we decided to hold a first exchange in the Victorian inn where Sue was hosted. Encouraged by the topic of the conference, I told Sue I had a preliminary question for her: "*What is an animal?*"

LAURENT. Do you know anything about it?

SUE *[laughing].* I have thought about it for years and years! A very long time. It is so clearly defined in the dictionary and everyone else seems to understand it so clearly. Would it even be legitimate to raise this question? By the time Kanzi and Panbanisha were grown, I couldn't possibly accept the definition that they were animals, and I couldn't stand it when people called them "animals" in their presence. They couldn't stand it either. It is one thing to ask "What is a person?" It is another to ask "What is an animal?" And, usually when you ask this question, people give you a list of animals.[4] Or, the animal is everything that is not a person. It is like asking what is day and what is night.

LAURENT. Yes.

SUE. People ask what is living and what is dead. I am not sure they have really asked before what is *animal* and what is *not* animal, except in regard to a plant. Am I right to say it is a question that you cannot ask?

LAURENT. It depends. In the pre-Christian West, "What is an animal?" was a fully acceptable question, at least in Greek or Roman thought. We can speculate that even prehistoric civilizations were raising the issue all the time. In his projects for a universal history and other associated writings, Georges Bataille made a great deal of Lascaux.[5] The cave paintings, in his view, often represent the reflexive moment of separation between humans and the other animals. It is true that in prehistoric paintings, one finds very few human faces. The emphasis is on geometrical signs, negative or positive (human) hands, and—above all—nonhuman animals. In contrast, in the Middle Ages or the Renaissance of the Western world, it is mainly the human face that is represented by painters. I would also say that the doubts about the degrees of separation between nonhuman animals and humans are clearly voiced in most human societies throughout history. We just happen to belong to a particular civilization with Judaic roots and certain philosophical and political institutions, a civilization where the question of human animality was no longer deemed to be extremely relevant—until it resurfaced relatively recently, because of biology especially. But then, in biological terms, it is difficult to really answer the question of what *is* an animal.

In ancient Greek, the word *zōon* (animal) literally means *something that is living.* For a Greek thinker, it would have been very difficult to completely exclude humans from the realm of "what is living." For us, the word *animal,* through its Latin etymology, is linked to the fact of being *animated.* It is not the same semantic spectrum. In ancient Greece, the main questions were "What is it to be living (and doomed to die)?" or "What is it to be mortal or immortal?" But, clearly, humans had to be *zōa* in the first place, that is, "living things" or "animals." Some ancient authors even proposed that there was a strong *resemblance* between humans and beasts that contemporary biology would not always deny. Overall, the distinction between humans and animals in classical Greece was not as vast as the one we both have been raised in. Even in a society whose gods were anthropomorphic and whose leading theoretical principles were anchored in rational protocols, even in a society that so greatly contributed to the theoretical shaping of our own, humans were largely considered to be very close to other animals—or even to be one "species," among others.

Now, if you worship gods that are partly animal, or if you believe in reincarnation and that, after your death, *your* soul could migrate into the body of a nonhuman animal, as is the case in many human civilizations of the past or the present, then the difference between *them* and *us* is overly frangible. Our current society inherits a relatively provincial view of what is a human—and of what is an animal. This view leads to the belief that "the animal" is a category that is very solidly constructed or even "natural"—which is inaccurate. You said that Kanzi and Panbanisha did not like to be called *animals.* As "humans," do we want to be called *animals,* or do we wish to reject this as well?

SUE. How we speak of ourselves is very important. It is particularly important when we are speaking around children, because we are shaping how they categorize the world: we may be limiting their views of how to even think about the world. Whether it may be a few philosophers and scientists who deal with these questions, most people today are going to go with them, based on whatever they heard when they were little.

LAURENT. Assuredly.

SUE. So, speech matters. And, of course I think there is a distinction between the living and the dead, as well as a distinction between the animated and the unanimated. Those follow the distinction between plant and animal. There is the distinction between those we believe to be immortal and those we don't believe to be such. Those considerations—while they are absent in today's biology—are extant throughout history and they remain in today's society. Most people would say that only humans are immortal—if they are going to deal with that issue at all.

LAURENT. In plenty of languages one can make a difference between an *animal* and a *beast.* The beast is the bad animal in many respects.

SUE. The concept of beast probably arose with domestication, with animals that humans can control and with animals that attack humans.

LAURENT. Yes.

SUE. But for all practical purposes, the *contemporary* issue surrounding the status of animals is what these different living beings are *under the law.* Under the current law of the United States of America, only humans can have rights, because only humans are supposed to have morality and reason, or to conceive of rights and, therefore, grant them. A human that no longer has brain function, or even a human born *without* brain function, still has rights that Kanzi and Panbanisha do not have. I suppose one would argue that humans have the *potential* for immortality, even when they don't have the potential to live a normal life down here. Therefore they deserve rights because of their human form. I don't know the laws of other societies very well. I know that Steven Wise has researched them through the Middle Ages and probably back further, and there were times when animals were brought to trial . . .[6] I don't know if they really tried to determine if the animal committed an act of aggression intentionally.

LAURENT. In ancient and medieval Europe especially, though not exclusively, secular trials or religious excommunications of animals were rather frequent. A very common case in the Western Middle Ages is about a pig eating a part of a newborn.

SUE. So the pig can go to trial for that?

LAURENT. Yes, in some instances. I also know that in some societies, including ancient Athens, an object could be sued. A knife could be sued and the knife would appear in court.

SUE. If a pig is eating part of a baby and they want to sue the pig, is this because the pig belongs to somebody else and they just can't kill it without affecting the livelihood of another person? . . .

LAURENT. There are cases where pigs as well as their owners are sued and condemned. The historian Michel

Pastoureau, who worked on this topic, speaks, for instance, of a late-fifteenth-century case, with a sow that is sentenced to be flayed for having devoured a baby, while its owners are sent on a pilgrimage so that they can ask the Virgin Mary for her forgiveness . . .[7]

SUE. If people were raising those animals for food, why would they have any compunction about killing them on the spot?

LAURENT. It turns out that pigs were also often left alone in the streets of villages and in the countryside, with little to no oversight, so they could appear to act "on their own." Pastoureau also stresses the medieval idea about pigs being the animal species closest to humans. In one famous instance in the fourteenth century, at Falaise, Normandy, the executioner even dresses a guilty sow with human clothing and puts on it the mask of a human, before torturing and hanging the animal. This being said, horses or worms could also be tried, and rats or crickets could be ordered to leave the place or risk excommunication.[8] At least one legal commentator from medieval France remarked that judging animals is absurd, because they have no "understanding."[9] In the Greek instance, objects were sued when the perpetrator was absent, or when there was no owner (as when a statue or a rock would hurt people by falling on them).[10]

SUE. Here, I think of two examples—there were many—that were very significant for the bonobos. First, Nyota always thought that tires had agency, that they could just roll and run over somebody, so he would go check them all the time: as if a tire were like a turtle sitting there and about to get up and move, becoming *animated* and having a will of its own. Tires also leave tracks on the ground like animals do. Nyota seemed to feel that way until he was about two and a half, when we tried to explain to him and show that tires were not alive. He just wouldn't have any of it! I have a second anecdote I guess I cited several times in papers. I was working with Kanzi, and I had a bunch of objects out, including a knife, and he had a bunch of objects out as well. I was preparing some test trial, and Kanzi was playing with his objects. He must have been eight or seven. He looked away because he was interested in what he was doing as I was cutting a piece of fruit for him. The knife slipped and it cut my hand. When Kanzi turned back around and saw my hand bleeding, he was stunned. He was like "What happened to you? How did this happen?" This was the question that was on his face, if you wish. He conveyed it by showing me my hand and looking at me with surprise, right in the eyes. I could have said, "Oh, I cut myself with the knife. It's okay." But I happened to be by a keyboard and I was at the time trying to use the lexigrams more systematically with Kanzi—it's always hard to use the keyboard, when you really just want to speak English and you know he can understand. So I didn't speak, and I very carefully said with the keyboard "KNIFE HURT SUE." Kanzi immediately picked the knife up and threw it across the room, because, in my keyboard use, and by means of grammar, I had treated the knife as an agent. I only did this because I didn't have the other words to put into the sentence. To me, this moment was the clearest example of the power of language, because Kanzi would not normally think of knives as agents. Through the fact of saying "KNIFE HURT SUE," the knife began to be treated as an animate being by Kanzi, as if it had jumped and hit me. Kanzi's reaction astonished me. I was already thinking he could understand syntax, but I didn't know the power that was vested in grammar and that it could be quite apart from the intentions of a speaker. This is an illustration of the ways in which language shapes everything we think, even when we don't intentionally want it to shape things in this way. Since the incident with Kanzi and the knife, I have reflected—I'm not sure I have been immensely successful, but . . . —I have reflected a great deal over my speech, over words or categories, and over grammar. In the studies I conducted with

Kanzi, when I would say a sentence and look specifically at what he did, or even when we were just in daily life, with the other bonobos, I would always explicitly look for a nonverbal way to recognize whether what I just said was understood the way I intended it to be.

LAURENT. You identified the fact that in our societies—in all societies—there are laws, and there might be a *legal* construction of what an animal is. Beside the legal, scientific, philosophical conceptions, there is a religious construct as well, that is being given to us from the day we are born, or even before this. There is also the narrative and fictional construction of "the animal." In the tales that we tell children, in fables, in literature, in movies, animals are supposed to do certain things and to be unable to do other things. The contours of what animals can and cannot do are very different—in our society at least—depending on the legal, religious, scientific, or fictional conceptions of "the animal." In the world we live in, for instance, monotheistic religions tend to put the animals far away from humans. The legal approach, as it now exists in the West and beyond, treats animals as particular objects, with added bifurcations between a pet and an animal, and between tame animals and "wildlife." But the category of the animal that is being constructed in fiction widely differs from the strict definition of the animal as a living object with very little—or no—agency.

SUE. That is absolutely true. Mickey Mouse is certainly an example of that! In cartoons and any kind of children's stories, animals become endowed with human capacities. On the other hand, if we want to dehumanize any group of humans, we simply give them animal capacities, making them *inferior*. In the symbolic constructions of the worlds, animals and humans can exchange places—humans can become animals and animals can become humans. In many young people today, there is an interest in zombies, something that isn't strictly human or animal and doesn't have conscious thought or reason, which is considered to be the dividing line between human and animal. But, of course, every character in *Winnie the Pooh* has consciousness and reason! Whether or not previous societies like those who painted the wonderful art in Lascaux—even though they often avoided representing their own face—whether or not they drew a distinction cannot, I think, be determined from the paintings themselves. Unless there is some written record and you know the kind of symbolic world that is constructed, the paintings can be interpreted in a variety of ways, including ways to reach altered states of mind.[11] There, you move into the religious domain and as soon as you move into something like religion or immortality, you are moving beyond this particular tangible physical plane, and moving there with your mind. Maybe you are moving there with the body, but if so, I think bodies become something, at that point, that functions in different ways than what we are talking about here!

When we ask "What is an animal?" I would take it that we are asking "What is an animal *thought to be* in our particular society today?" One could make a distinction between humans and animals on the basis of immortality or soul, but since humans can't be legitimately excluded out of the animal realm if you are a scientist, then another list of categories—like reason, consciousness, self-consciousness, ability to plan ahead, and, of course, language, culture, and tools—would be alleged to justify the divide. For instance, one could take *reason,* which is difficult to define but could be operationalized into a scientific paradigm, then ask: "Is a nonhuman animal capable of reason or not?" Some scientists would give a positive answer, and others a negative one. From there, one could alter the topic slightly, moving from *reasoning* to *thinking,* and one could define thinking a little bit differently than one would define reasoning. And this could generate a lot more questions. But, quite frankly, if you are discussing with somebody who might own four dogs, some parrots and cats, and maybe an odd exotic animal in her home, or who reared animals all her life—or if you are exchanging with somebody who comes to visit those animals, who has never owned any, and eats meat, or if you are talking with somebody who is vegan—then there is almost no disagreement on the practical existence of the divide. All those people will probably say, "Oh yes, these are animals and these are humans." You get very little disagreement on the *definitional* issue. Now, if, among that same group of people, you ask "Do animals have souls?" you might get a lot of differences. My sister Liz once told me that Clara, who was Congolese and a very religious person who used to work with the bonobos, certainly treated them exactly as people. She was very good, and I never had any discussions with her about whether apes were animals: it did not seem to be a problem for her. Liz and Clara had long conversations about whether or not Panbanisha had a soul, right in front of Panbanisha . . .

While those discussions are not what constitutes a body of science, or a body of literature, the scientists and the philosophers that grow up raising such questions come out from whatever society has raised them. They come to their adult discussions with certain thoughts, feelings, and categories in mind, and when they reason, they reason based either on how the majority of people would use those categories, or on some data and paradigm that they think separate humans and animals. Thus, the basic question is: Do we wish to separate humans and animals, because we are egocentric and because we are the humans? Each human group thinks of itself as "the Humans," and those other groups are either *lesser* or simply *nonhuman* animals. Those other groups will then say "No! we *are* the Humans, and the other ones on the other side of the valley are lesser than we are, they are subhuman in some way, if not animal." Is our need to do this an egocentric one? Or are we really trying to grasp something that we feel is fundamentally different between humans and nonhumans? What is at the basis of this category? Why does it exist? A group calling itself "the humans" is a self-reproducing group: its members want to continue, and they feel the other group might attack them, so they want to self-identify as the group of people that are going to fight together. I would imagine that since chimpanzees exist in a very similar state and face chimpanzee individuals across the valley that could come kill them, they would probably call themselves "the Chimpanzees!" I don't think bonobos would do this. By all reports, they don't. Intellectually, it is just very clear that chimpanzees are capable of entertaining such a thought, and the dual fact that chimpanzees treat other chimpanzees this way, and bonobos don't, certainly raises questions of how bonobos categorize each other and the world. Do their categorizations of the world cause them to treat other bonobos differently? Chimpanzees generally don't kill and eat other chimpanzees. When they do, the record is that they are very, very vicious. Chimpanzees rip testicles off, they rip the faces off, they do all kinds of things that, when they decide to kill an enemy chimpanzee, are not really necessary. There has been no record of bonobos showing that behavior. There are human societies that don't do it and, of course, there are many human societies that do it. Chimpanzees kill baboons all the time and they kill other kinds of monkeys, and they eat them. Is there a basic distinction in the chimpanzee mind between *chimpanzees* and *animals*? Could there possibly be such a distinction in a baboon mind—"I'm a baboon and these are other animals"? I don't think a baboon could conceive of it.

LAURENT. The categorical discrepancy has a lot to do with the question of the *imaginary* of the group that is built through the image of the self or the recognition of the other as a similar other, with the ability to voice those differences. Suddenly, everything that is discursive is strengthening or modifying the categories between *us* and *non-us,* in terms of what is living. There is no reason to consider that discourse would be the only basis for *producing* the difference between us and non-us. If a nonverbal primate is able to identify the other as a similar other, as it happens with chimpanzees, then one could certainly admit there is something like a descriptive category of in-group members. If we take your example of chimpanzees eating baboons but very rarely (or never) adult chimpanzees, then we encounter the issue of cannibalism. Most human societies where cannibalism was practiced in "modern" times were heavily stigmatized. The records that we have and the ethnological literature about it are usually biased. In some instances, there is an exception to disapproval, as when Montaigne argued in his *Essays* that "Cannibals" tend to be at least as civilized as Europeans are. Despite this minority opinion, the main approach was to see cannibalism as something unacceptable, and, in historical times, the practice really receded, virtually vanishing. But, as far as we can tell, in some of the societies where cannibalism was being practiced, the idea of *becoming* the other through the act of eating the other was often key.

SUE. What I was trying to say in a much-reduced way, about chimpanzees and baboons, is that chimpanzees hunt baboons for food.

LAURENT. Yes.

SUE. When humans engage in cannibalism, it is often for the kinds of symbolic eating that you talk about. In fact, there are special plates, special knives, and special forks for this, and sometimes only the person that killed the other person can share that particular human. Except in extreme circumstances where people are extraordinarily starved, cannibalism has nothing to do with hunger.[12] And even when hunger is the cause, there is no *hunt*. People who engaged in this kind of "survival cannibalism" would even do things like draw straws to see who will die and be eaten. Chimpanzees do not hunt other chimpanzees for food, although they hunt other primates for food. Would chimpanzees ever develop a ritualistic, symbolic practice of cannibalism? This, I think, is a very important question. The fact is that the only chimpanzees they eat are babies—who do not have autobiographical selves, who do not have personalities, who do not have histories, who do not clearly show that they know who they are, something that happens much later than in human babies. The human practice of cannibalism is heavily endowed with symbolic causes. You are ingesting the spirit of this other warrior, and the way to do it is through eating his flesh: you are not ingesting the flesh to have the flesh—you are ingesting the flesh to have the spirit. The spirit is intangible. You can't see it, you can't point to it. It doesn't exist in this plane, it only exists in the mind. This is why Jane Goodall's early reports, which I think she backed off of, that chimpanzees were doing rain dances, made such a splash.[13] Because chimpanzees were seen as reacting to something intangible, to something like a god. What reason, or consciousness, or the symbolic process allows you to do is to create an imaginary world. This was one of the main reasons why I created the "Mythology," with Gorilla, Bunny, Pinky, and the Rain Monster.[14] Each character had a personality, and they each did different things. They began to have an autobiographical history. I wanted to see if a symbolic world was meaningful to Kanzi and Panbanisha. I found out that not only was it meaningful, it was *immensely* so, and it could even have more power than the regular world. It was so immensely powerful that I became hesitant to deal with it, and other people recommended that I dismantle it, which I eventually did. In retrospect, I don't really think this was a good idea. One creates concepts, thoughts, desires, wishes, and entire roles by saying things like "KNIFE HURT SUE." By beginning to play around with what GORILLA was going to do, or what GORILLA was, or what GORILLA thought, or with a story about GORILLA looking for BUNNY, one instantaneously had Kanzi's and Panbanisha's full attention. We could make videos of Gorilla and Bunny; we'd also write stories about them. Panbanisha and Kanzi could understand. The video and the costumes were just a tiny thing to solidify the Mythology, but the *reality* of the Mythology lay in the imaginary and the symbolic.

All this rapidly made it very difficult for me to think of Kanzi and Panbanisha as animals. What was happening with them was not something I could do with dogs, or with cats. Could it be done with any organism that recognizes a red dot on itself in a mirror? I don't think so, because all this happened through the vehicle of language. More recently, I came to hypothesize that one of the reasons that humanity spread so rapidly around the globe is because of its ability to create these symbolic worlds. We can create a symbolic world that corrects all the inequities of our current world, of our current little village in our current tribe, so we can get eight or nine people to agree with us and we just go, we leave to create the ultimate society, whether it is a commune or a democracy, or another place that somebody in particular wants to be the king of. We can leave and go create a "new world." We have been doing this for some time, and we have basically ended up colonizing the whole globe. The task is getting harder now and we have to communicate more with each other.

Why haven't apes done the same, if they have a symbolic capacity? Kanzi and Panbanisha entered the symbolic world very easily. The next set of problems was, as with humans, the degree of reality that the symbolic could reach, the boundary between the symbolic and the real, and the ability to distinguish between the two. It was very easy for Panbanisha to differentiate. Kanzi sometimes gets caught in between the two. Nyota does not get caught between the two. Possibly, Kanzi's early rearing by Matata made a difference. As I began to create those symbolic worlds and see the variations in the bonobos themselves, I knew I was not dealing with anything like "bonoboness." I was dealing with language and concepts. Matata was always very interested in the symbolic worlds, but she always seemed to have a very different take on them than Kanzi and Panbanisha had. Matata always acted as though she knew there were alternative realities and as though she were very familiar with this concept, far more than I was. I first realized it when I was working at the Yerkes Center. Matata was very young, maybe seven or eight, and I picked up a stick that happened to be black. I was twirling it around me, and she looked at me with a wary face. As if the stick had supernatural properties and I could act like I had powers. She would immediately try to steal that particular stick from me. Much later, when I went to Wamba, Congo, I was told by the local people that they thought bonobos had an understanding of the supernatural, and that the bonobos endowed certain objects with supernatural properties. The people of Wamba used the term "brothers" for bonobos and they did not use that term for any other animal. They attributed human capabilities to the bonobos, precisely because they thought the bonobos had *supernatural* capabilities—and not because of any issue about reasoning, consciousness, the ability to use tools, or any of those things that the society from which I came would judge them as being animals, or nonanimals. The people from Wamba were certain that the bonobos were capable of supernatural feats, hence of symbolic activities. Therefore they included them into the category of human, as "brothers."

As we try to put all those pieces together and ask "What is an animal?" one answer could be, I believe: an animal is a being that does not have the capacities that we recognize in our own species. As scientists, we raise the question in a very pedantic matter and we make an assumption that no other species has the abilities we have. Contrary to many other societies, our particular culture begins with that assumption.

LAURENT. I will try to add another point on cannibalism. The reactions of the European explorers to the "New World" when they discovered it—or rather claimed they *discovered* the practice—was that it was absolutely inhuman. Precisely no, it is eminently human.

SUE. Here, I have a question I have to ask.

LAURENT. Go ahead.

SUE. The thought that immediately occurred to me is that, in every instance of cannibalism we know of, there also is fire. I haven't heard of any organized raw cannibalism. Is fire transforming the flesh in a way that makes it symbolically acceptable to eat? The existence and control of fire is a very defining characteristic separating humans and other beings on this planet, whether you want to call them animals or not. Kanzi's earliest fascination with video was for the movie *Quest for Fire*. He must have watched it a thousand times. I don't think we can talk about cannibalism per se without talking about the physicality of how one is actually going to eat another person. It seems to me too theoretically divorced. If you cooked a chimpanzee and offered it to another chimpanzee, it might eat it. But if you don't have fire . . . I suppose that one could eat humans raw as soon as one develops the symbolic concept of incorporating the essence of a being by devouring it.

LAURENT. The same is happening with regular meats: one also incorporates the symbolic force, or power, coming from an animal.

SUE. Either way, it is symbolic and my question is, Do chimpanzees have that same kind of symbolic concept about that which they eat? The moment one debates what is human and what is animal, at some very basic level, one has to examine what a symbolic mind is and what it is not, what a symbolic mind does and does not do. We are getting pretty much to the core of the issue when we start talking about food and symbolism. The other interesting observation that has come out of the wild in this regard is chimpanzees eating medicinal plants.[15] Is there any symbolic ability in there? This is undoubtedly the case in humans when they eat medicinal plants. Are chimpanzees merely noting that they sometimes feel better when they eat those plants? Or do they have a whole medicine-like concept? Let me cite another example from the bonobos. As you know, Matata was not acquiring our symbol system. I say it with memories of sitting with her at the keyboard, for maybe two or three years in a row, day after day, hours and hours, trying to teach Matata six symbols and later seeing Kanzi and Panbanisha learning them almost by magic. There was just a huge difference! You have to ask if that capability was *in* Kanzi and Panbanisha. Does it only exist in a baby? I could understand that it might be more difficult for an adult, as it is always more difficult when people change cultures to learn new language and customs—but they do adapt. It does not take three years of sitting there with six symbols. I always had the impression that Matata didn't really like our language and she really wasn't about learning it. She thought it was ridiculous to punch symbols on a board. Why didn't you just speak with your mouth and speak bonobo like she did? She would sometimes point to things and utter sounds, or point out to the forest and utter sounds, as though I should be able to understand her—but I couldn't. It just seemed to me she had a very high disregard for my inability to acquire her language, in the same sense that many humans have a very high disregard for chimpanzees, if they think they can't acquire language the way human children can. As the whole group began living together more in Iowa and was often housed in the same building, Matata could be caught studying the keyboard when nobody was looking, and she started using some symbols. Elikya and Maisha also started using the symbols and nobody taught them—unless Kanzi and Panbanisha were teaching them in secret. It looks as if there was some hidden competency they had.

Matata was once gravely ill. She went to the keyboard and looked directly at me and used the lexigrams "GIVE GREEN MEDICINE" in a sequence. I had no idea that she even knew any of those symbols, but the fact she uttered the sentence while being sick, in such a highly ordered fashion and looking right at me, suggests that she did know them. She was not randomly looking around the keyboard. She went right up there and said it. Of course, in the wild, if Matata were sick, and if bonobos have some competency of botany, she might know exactly which green plant she needs. Because scientists have looked at apes with the eyes of outside observers, they have drawn simple conclusions about primate behavior. We are very much in the position of what people thought of human primitive societies before anthropologists decided to become a part of the group and learned the symbolic structures, the kinship structures, and the language of the others. Until we do this or at least until somebody makes a serious attempt to do it with chimpanzees, bonobos, gorillas, and orangutans, we'll have no idea of what is really going on in them. I don't know whether I would be willing to make a similar statement about macaques or baboons or other nonhumans. This is why the whole category of animal seems a bit fuzzy to me at the moment.

LAURENT. We can find a strong connection between the abilities to create imaginary worlds (or beings) and to speak about them, to make them alive. Everything that pertains to imagination and to narratives where many nonhuman animals are talking, singing, dancing, building houses, etc., reveals that "we" only have an operational category of what an animal is. It is *operational* in the sense that the "obvious" concept works, but it is inextricably impossible to explain, and possibly disputable through means of argumentation, speculation, or fiction. This category, in our society, is strengthened in a very decisive manner by our legal institutions, by our religious habits, by scientific knowledge, etc. Then, when it comes to the imaginary, we are suddenly deconstructing de facto the whole category. By having animals that dance, sing, and converse with us, we are just undoing the category that is otherwise assumed to be "self-evident." This is closely tied to the nature of fiction. Fiction is routinely described as something that is not real but that people take for real. It is much more complicated, since in many cases, in the arts especially, you process fiction as something both fictitious and real. Human children, when they are younger, have a harder time addressing this, because they tend to switch from one side to the other. As you described him, Kanzi could sometimes be caught by the realness of fiction, while concurrently losing sight of the fictionality of fiction. As for all of our stories and pictures about animals, we know them to be real, and we know them to be fictitious. We know that there is no real set of definitional categories or qualities that would absolutely put us outside of any animal category. Narratives remind us that the distinction between human and nonhuman animals is, in itself, a fiction. But it is not any fiction: *it is a fiction adjudicating fictionality,* and through which "we," symbolically and practically, recognize ourselves as being nonanimals.[16]

SUE. What human children spend most of their time doing, from the moment they are two, is imaginary play. They lift toys around and give them properties, or they play with their policemen or nurses. They also sometimes create imaginary playmates—not all children, but many of them. They treat their imaginary playmates as being completely real. If you discuss the matter with them, the questions you ask seem to frustrate them. They will usually try to respond, as though the questions reached some mental distinction they have between an imaginary and a real playmate, but they never really answer those questions very well. What they eventually do is get rid of the imaginary playmate, to conform with adult views, I guess. But certainly, as you point out, they have no trouble giving those characteristics to animals. The whole animal-rights movement is a function of many of those early children's stories and cartoons. Some individuals, as they grow up, don't feel comfortable with the societal disattribution of those characteristics to animals. In movies like *Babe,* animals retain those features. Kanzi and Panbanisha just loved that film and they were quite happy thinking that Babe had all those characteristics. Many adults feel that pigs have them. They may understand that pigs don't *actually* speak English, but they don't really want to take those abilities away from animals. Many participants in the animal-rights movement use the chimpanzee, and Goodall's work, as an example—but they would quite graciously extend capabilities a long way down the animal scale.[17]

Then one has to ask: Why, in our society, do we give children one set of criteria for animals with all these capabilities, and why, as adults, do we feel a need to take those things away from them? Shouldn't we expect that, if we do this en masse, the whole issue of animal rights is bound to arise, especially as we move away from the small farming communities—where people actually see animals alive and experience animals being killed and eaten—to factory farms that are hidden away and deliver you a package of meat that has no relationship to the animal per se? It seems to me that we are fostering a strange kind of dissociation between what one can do as a child and what one is expected to do as an adult.

Most people do feel that animals have emotions and therefore we'd like to avoid thinking that we are actually eating them. So we disguise them. Not only do we put them in plastic packages and never see a whole animal, we disguise them with all kinds of other things. It is just a little part of something that we buy that has protein in it, and it just happens to be animal protein. Our society seems to have no problem telling adults that animals are not really symbolic.

LAURENT. More and more, these imaginary stories or pictorial representations of animals having human traits tend to be reserved to childhood in our societies, but that was not the case across centuries, and that is still not the case in many civilizations. One of the paradoxes is that, in very rural societies where people are actually in touch with plenty of other animals, instead of just knowing their pets, one usually finds much more stories and fables about talking animals. Those stories or pictures are as much for adults as they are for children. We have witnessed in the last few decades a strong restriction on the imaginary, with the idea that an artwork or a narrative that would challenge the boundaries between humans and animals is more suitable to children. In the wake of your own comments, I would be tempted to say that, within the animal-rights movement, the performative fiction of the law is seen as the acceptable continuation of children's books and movies about talking animals.

I also believe that the highly ritualistic ways animals were being killed across millennia for food consumption and/or sacrifice were leaving little doubt about what humans were doing. Nobody could ignore that a living animal had been killed and then eaten. Whereas our present situation with farm industries and with animals being raised and killed in places nobody can ever watch is, of course, a manner of obfuscating the *symbolic* killing of animals. As the symbolic is lacking, we are reinjecting it, thanks to a universal appeal to transspecies communication through pain—and to the organized enrollment of childhood memories.

SUE. At the simplest level, an animal in our world today is de facto a living, nonvegetal organism that you could raise or hunt, and eat.[18] Of course, there are individual likes and dislikes, as well as religious or traditional interdictions on some food consumptions . . .

LAURENT. . . . Then there is this distinction between animals and pets, and all the rage that you can now see on the global scene against people who eat dogs or horses.

SUE. I am aware that you feel differently about domestic animals or pets. But you don't have anything like cannibalism appearing to *enable* you to eat a dog or a horse. In certain countries, all kinds of fish are appropriate to eat while no one would eat some fish in other countries. In some places, eating insect is part of gastronomy, and elsewhere this would be considered to be appalling. There are certain foods, even plants, that we eat and don't eat, and fish that we eat and don't eat. What you eat is very cultural. But while we can consume animals, with cultural exceptions, we don't *consume* humans. Cannibals do not hunt or farm humans for food. Humans do have a knowledge of who they are that is engendered by their symbolic mind. A symbolic mind allows a human to stand out of the ongoing reality and think about the human as something *other* than what the human is doing. Nonhuman animals may be able to plan ahead, they may be able to remember, but they don't seem to stand apart from their bodies and think about themselves as apart from what they are doing. Humans have such a capacity. It is this capacity that rides on language, unless language rides on it. I am not exactly sure I know the answer to the question at this point. Then, there is second-order thinking.[19] And humans are capable of third-, fourth-, and fifth-order thinking. If one eats a living being that has a meta-mind or a symbolic mind, one will need to justify it by other kinds of symbolic practices. While the human mind is able to justify any behavior it wishes, the lack of organized farming for the consumption of humans as food is striking.

LAURENT. All this would leave us with transient and different answers to our initial question. An animal is a dissimilar other that could be farmed, hunted, eaten, consumed. This category relies on group dynamics, and, mutatis mutandis, it could exist analogically in some other species. The conceptual divergence between "humans" and "animals" is not completely tenable—even though it holds. It holds through social, political, religious, legal institutions. It is additionally justified, shaped, and perhaps tangentially undone by philosophical, scientific, and literary means. In this, the category of the animal is like many other fundamental aspects of our symbolic life: it is an operative, sometimes performative, fiction. But the very possibility of entertaining such real fictions is what is at stake in the metadefinitional aspect of "humanness" and "animalness." By pushing forward the fiction of a divide among animals (whatever this divide could be), we seek to humanize ourselves. We produce ourselves as those super-natural animals, as beings that claim to absent themselves from the given organization of the real, from *phusis* itself. Being a human, then, is a particular *process* emerging from the ongoing development of the symbolic meta-mind. It is not a durable attribute. It is never granted once and for all—except in a legal or political fiction. The different ways of framing the animal–human divide coexist, but they do not have to coincide. Undoubtedly, coincidence is sought after, as when new scientific results are being mentioned in court about the issue of ape personhood, or when politicians promulgate a welfare law based on philosophical debates about pain. Just, perfect coincidence is out of our reach. In other words, we do not know what an animal is, though we are quite good at doing as if we knew.

SUE. I guess we reached the end of our first dialogue, but may I add a few words on what it is to be an ape?

LAURENT. Yes, by all means.

SUE. There is a scientific definition of what is an ape, and it has to do with anatomy, then with genetics. There is also the consideration of how, in the past, the definition of a species has been made. It may be time to change the definition of what is an ape and of what is a species, because our former categories are being challenged by new scientific findings all the time. But another issue is: Is there a need for such categories? Why do those categories exist? Are they factually constructed as a way of easing our thinking? Or do they really exist at a deeper level? Knowing what is an ape has something to do with what is human, and what is not. In the scientific sense, both apes and humans are classified as primates. But humans are said to belong to a *special* family called *Homo*—a family no other ape can ever be a part of, no matter whether it uses tools or language. Even though both chimpanzees and bonobos are genetically closer to humans than to other apes, there is today no consensus to support their inclusion into the *Homo* genus.[20] This highlights that we want a special designation for ourselves: we can be an ape, all right, but we have to be a different kind of ape. In my view, this has to do with the traditional theological literature assuring that humans are the only beings with a soul.

LAURENT. Carl Linnaeus, the inventor of taxonomy, classifies orangutans as the "troglodyte" members of *Homo*. He calls them *Homo nocturnus* (their alias being *Homo sylvestris,* Bondt's Latin translation of the Malay *orang utan,* "man from the forest"). Linnaeus writes that *Homo nocturnus* "thinks" and "speaks by whistling," a statement that Buffon will ridicule.[21] Buffon puts orangutans and chimpanzees outside of *Homo,* in arguing that apes, because of their anatomy, cannot have fluent speech. (Obviously, verbal fluency is not the same thing as whistling, but the argument stuck.) At the very beginning of scientific taxonomy, we find both the theoretical possibility of including some great apes into the *Homo* genus—and the rejection of this idea, based on the anatomy of language.

SUE. Then, why is so much attention given to speech, or to the word, or to language? Because language is seen as the defining characteristic that would keep apes out of *Homo*. Whenever people actually tried to teach apes to speak, they got no convincing results, as could be seen in the work of the Hayes with Viki. The Gardners, David Premack, and Duane Rumbaugh got around it by other methods (a variation on American sign language, a series of lexigrams). The apes were suddenly said to have language—yet, they still didn't speak. One of the things I did was to attach a voice to a keyboard. A lot of antagonism came because speech was now attached to the keyboard! . . . I am being told that, in the Ape Cognition and Conservation Initiative (ACCI) laboratory in Des Moines, where the bonobos I have raised are kept at the moment we are having this dialogue, the speaking keyboards are no longer used in daily life: this a way of making the bonobos "voiceless." Some time ago, the persons in charge have apparently spread the word that they want to "put the bonobo back" into the bonobos. One does it by taking away their language.

The basic consensus is that humans can speak and apes are silent. Apes are categorically *silenced,* if you want. Now, take any human child who is autistic or retarded. The main thing teachers and parents wish is to have a child that speaks. More than anything else. Years ago, when we began working on a companion project between autistic or retarded children and Kanzi, I conveyed that the goal was not to teach these children to speak. It was about establishing an environment in which speech would become traded between individuals as a part of behavior. You don't sit children down and teach them to speak. You incorporate them into your life through speech. At that point, in the 1980s, we had the first "talking" keyboards. How much the parents loved those keyboards! No one in the autistic or retarded community who had previously worked with us ever loved the silent keyboards that Lana, or Sherman and Austin, had previously used. Human parents thought of those devices as mere didactic tools. This all changed when, by pressing a key, a synthesized word could be uttered by a computer. We had designed a routine. One would begin an interaction with CHASE or TICKLE. You could say CHASE, and you could play chase; you could say TICKLE, and you could tickle. You could make manifest what a TICKLE was. You could make manifest what a HAT was: you could press the HAT key, and you could put a hat on. It was now easy to make certain overt actions, if you wanted to use the symbol. But let us say that you want to use a symbol like IS. What is "IS"? "IS" is "is." If you are trying to teach an ape that you assume has no language, and you utter the sentence "What is 'IS'? 'IS' is 'is,'" then what have you said? It means nothing, unless one already understands the concept of grammar, which involves the understanding of the verb *to be*. And if one already possesses all that is needed to understand "IS," how did one beget it? Such was the mystery of language I was first trying to face with the chimpanzees Sherman and Austin in the 1970s. It seemed Kanzi solved the problem on his own.

When we began to work with parents of nonverbal children, what they mainly wanted was their kids learning PLEASE, THANK YOU, HELLO, and HOW ARE YOU? They wanted their kids to be able to take their talking keyboards to go to McDonald's or the grocery store. This experience dramatically emphasized the discrepancy between what I was calling language as a primatologist wishing to report results to other scientists, and what language meant to those parents. How could I push an ape around town in a wheelchair and have it say "HELLO HOW ARE YOU?" to the people at McDonald's? Even if it were good for the ape, necessary for the study of language, and very helpful to me, I couldn't possibly do this! So how could I conduct the research that was needed to deliver what the parents wanted? This is where the two programs began to split—because apes had to live in one culture, and human children had to live in another culture. The cultural expectancies of the human parents were driving them in a direction that was not the one the scientific world was making me favor. However, all this was still "language." It was still "culture." The dialogues between the scientists working with the children and the scientists working with the apes began to collapse. They had once been very good; there was nothing wrong personally among the people who were working in those two fields. It was like an ideal marriage in a sense. But the problem hit with the speaking keyboards and what the parents wanted them for, compared with what I was allowed to do with apes. Say I invented a forest where one could go, a forest with inhabitants. Instead of finding a pineapple in the forest, a bonobo would find someone who wants the ape to come up with its speaking keyboard and say "HOW ARE YOU TODAY?" Say that I designed this experimental setting, and that Kanzi could do that. How would I prove to scientists that Kanzi *knew* what "HOW ARE YOU?" even *meant*? What would I do?

When I talked about this issue to Stuart Shanker and Talbot Taylor, who were working with me at the time, Talbot began focusing explicitly on the word SORRY. We wrote back and forth for several years about this. Would Kanzi be able to understand the word SORRY, and if he were, would he *really* be sorry? If he were really sorry, how would I know he was? This discussion went round and round. The parents of autistic children wanted SORRY to appear on the keyboard, because if the kid did something they didn't like, they would teach him or her to say "SORRY." Nobody cared if the kid really knew what SORRY meant or if the child were really sorry. They didn't even want you to address such questions. They just wanted the child to be able to say to another person that he or she was sorry, so that the other person could carry on—so that, simply put, the whole world could carry on. In science, the world couldn't carry on, because we never knew if the "inner" Kanzi was actually sorry.

The whole idea of what is language and what is speech was there connected to preconceptions about species. Because a child is human and can say SORRY, the whole community accepts that the child is sorry. Since she said it. If the child does a bad thing over and over again, one might say, "Well, you were not really sorry," but the utterance will still be taken at face value. Working with bonobos, I had to begin a philosophical discussion to grant the slightest standing to what the ape was really doing and to my own claims about the ape. I even had to go into the subjective state of the ape, which seemed a literal impossibility. I realized that humanness is defined by what humans dialogically accept, as we co-construct

> our cultural identity. To be an ape in a scientific study is not dialogical, it is largely predetermined. This was very frustrating to Kanzi; as he grew older, he began to get a sense of it. I think Panbanisha got a very deep sense that "being an ape" in the environment she lived in was, more than anything else, about being evaluated and studied by members of another species. She couldn't just *be her*: she had to *be an ape*. Everything that she was had to be evaluated by this other species, because Liz, myself, and others had to constantly address those questions. She understood it by looking at us videotaping her, or when TV crews came, like NBC or NHK. She saw what we had to do in order to try to convince other people that she could actually understand language. I finally concluded that humans need the category of something that is *not* human in order to define themselves, and what they are. All the unique properties we have assigned to ourselves, none of them work anymore. So we need to know what we are not and, within the "animal" realm, we *want* apes to embody that which we are not and that which we would prefer not to be.

The evening lights of the inn were on. I could hear a slight humming sound in the parlor room. Our preliminary dialogue had run its course. It had opened up the different paths we would later explore.

On Dialogue and Consciousness

Until 2004, Kanzi, Panbanisha, and Nyota had lived their entire lives in Atlanta, at the Georgia State University's Language Research Center, where they had a forest to explore. The move to Iowa promised to be an improvement, with a new and larger building, the possibility for bringing other ape species, an outdoor area, many staff members, and the financial support of a philanthropist. However, as I was visiting the laboratory for a second time, some six years after the transfer, and after the benefactor had announced that he would stop funding the lab, the Great Ape Trust already looked like a deserted fortress in the marsh. Yet, compared with the preceding year, it was now possible for the apes to at least go into the yard, which had not been the case since the 2008 flood. Teco's progress was outstanding. And Kanzi had decided to make fire again. At the same time I was at the Trust in 2011, a reporter from a British media agency came to photograph and videotape Kanzi Prometheus. The London office of the agency had called Sue several days in advance, with a specific request: they did not want Kanzi's genitalia to be visible on the pictures. After some back and forth, Sue had to accept the principle to cover Kanzi's private parts. She decided to sew a skirt, so she and I both went to Walmart to buy a yard of fabric. I suggested that a fig leaf could have been another option, but we were not sure to find one of the right size in Iowa. At the supermarket, I asked Sue if she really thought she would sew a skirt before the arrival of the photographer the next morning. She answered: "You're right, I won't do it." We opted for buying briefs instead. There was an irresistible sale on superhero underwear: we went back to the lab with Green Lantern, Batman, and

Superman briefs. “Ask Kanzi which ones he prefers,” Sue told me, as she was going to the other side of the building. Here I was, talking in English to Kanzi, who was perched on its platform behind the mesh. “Kanzi, you must wear briefs for tomorrow’s shooting, so tell me the ones you’d like best,” I said. I wasn’t sure how Kanzi would signify his preference to me, but it quickly appeared that head signs were the easiest way. I first showed the Batman underwear; it was immediately rejected by Kanzi, who shook his head energetically. The second option was the right one: Green Lantern was acclaimed by nodding. The Superman underpants no longer had any interest for the fire maker.

The time was now 2014. I was in Des Moines, but not staying for long. Sue picked me up at my hotel and drove me to the house that she had been renting for a few years. It sits next to the lab. Sue really wanted us to discuss “consciousness.” I may have begun our second dialogue like this.

LAURENT. Much more than the concepts of *language, animal,* or *free will,* the concept of *consciousness* is very “intermittent” in philosophy. While it is possibly true that medieval philosophers were less interested in what is an animal than the ancient Greeks were, or that language was sometimes downplayed before becoming paramount in the twentieth century, even contributing to the rift between philosophy and its “analytic” offshoot—there remains that the history of the *concept* of consciousness is full of interruptions. Consciousness is largely a non-problem before the Renaissance, or it would belong to series of subaltern technical considerations (on sleep, for instance). This is a first difficulty: if consciousness is what it is supposed to be and the basis for all we are and do, how is it that its concept would, as such, be barely mentioned or constructed, over centuries of philosophical reflection? Why does consciousness suddenly matter so much in many doctrines, especially from the Romantic era onward? Of course, this “as such” is a hurdle in its own turn, for the *as such* relates to what we tend to mean when we use the term “consciousness.” There arises a second difficulty, linked to the previous one. As consciousness became this key concept, it has been challenged through an emphasis on the unconscious. We could say that the two problems I have just located are, in fact, historical contingencies: it "just happens" that consciousness is a marginal question for centuries, or that writing about consciousness would open up reflections on the power of the unconscious. I rather believe this conceptual intermittence and this inner challenge to be parts of what would constitute the theoretical question of "consciousness."

Do we need a concept of consciousness? Can we do without it? In principle, any philosophical concept that has been created then segmented and aggregated could be critiqued or even "eliminated." It is always possible to say "There is no such thing as language," and add, for instance, "There are only languages, or even idioms, and maybe speech and discourse, but language—as such—doesn't exist." The same line of argument could be applied to anything that is constructed: you can use it for "society," "madness," "humanness," etc. But since everything we *think* is at least minimally constructed by the fact of its thinking, this sort of argumentation is both very radical and rather trivial. It would sound quite paradoxical to deny the existence of language—through the use of language. In my view, though, the case of consciousness is different. I would say that one can develop an entire philosophy of mind without speaking too much of *conscious* or *unconscious* operations. I am not saying this would be right, but I guess that, in my *Intellective Space,* I did not use the words *conscious* or *unconscious* so much—so I am not saying this would be wrong either . . .

SUE. True, anytime you take a concept like "language," "consciousness," or "soul," this always raises the question of how everybody is able to talk about those things and use those words with some degree of sophistication, while seeming to understand them. There is the shared feeling of understanding some of the essence of those concepts or objects, without having to come to some formal agreement on them. Well, if you are not arriving at a formal agreement, then what kind of agreement have you arrived at through your dialogue with others that would give you any sense that you "know" what these words mean to you in some sense? Meaning can't arise from a definition, even though meaning is arising. Maybe by just talking of nebulous terms like "soul," "mind," "consciousness," we are really touching on the more complicated areas of meaning. Meaning is something that I have no way at all to build, unless I engage in a kind of—I don't know how you would call it—a kind of dialogue with oneself . . .

LAURENT. This is how Plato defines thought: "a dialogue of the mind with itself."[1]

SUE. I can't imagine that, if I am in a dialogue with the mind and myself, meaning would be a big issue. But if I am in a dialogue with somebody else, meaning definitely is! "Do you understand what *I mean* when I say *x, y,* or *z*?" I know what "I" *mean.* I know what I mean by *consciousness.* I know what I mean by *soul,* by *thought,* by *mind.* I have my own meaning for those things; I just don't really know how I arrived at it. If I am going to have a dialogue with you, with the readers of the book, or with anyone else, there is some necessary attempt to figure out what I mean and to have a correlation with what you mean. We take other words and we attach them to the words *meaning, soul,* or *consciousness,* trying to enable any future self-dialogue we would have about them to exist in some common conceptual space. The aim is to reach a level of sufficient agreement. Now, we have no problem with a word like *golf club* or *shoe.*

I submit that the only way we could attach any meaning to a word like *soul, consciousness,* or *thought* is when we talk about their being lacking: when we see something happening and say "She is unconscious," "He seems to have lost his soul," or "They are not thinking." We recognize an abnormality. We are able to apply those terms in the *absence* of a behavior that allows us to consider the normal behavior. Moreover, because we can talk about a person having a mind or a soul, a person being able to think and have consciousness *all at the same time,* this gives us the impression—and illusion—that those are all different qualities that one can possess. I am naming some internal qualities that I possess—that my consciousness is not the same as my mind, which is not the same as my soul—and the only way that I can attach differential feelings, ideas, or thoughts to them is by trying to compartmentalize myself and to attach certain ideas about myself to each of those different words. How does this semantic process happen? When it happens, is it happening in the same way in you and in somebody else? This process is probably a function of the linguistic history of the term, of what you have heard or read people say.

Now, *consciousness* and *unconsciousness* may be a little clearer, because there are times when everyone goes to sleep. Your body is not doing the same types of things as it is doing when it is not asleep. If you are asleep and I can't talk to you, I can think of you as being unconscious. I can call you unconscious. But, in your sleeping state, you might be quite conscious and thinking of many different things. How would you even know if you weren't conscious, when you are asleep and a machine can register you as not dreaming, as having no perception of being unconscious? I don't see any real direct answer to such questions, except by saying that animals are not conscious in the same way that *we* are. They don't think the same we do, and their thoughts are not rational and ordered in the same way our thoughts are rational and ordered. Even if you think animals might have some thoughts, perhaps even a soul, and that they are conscious because they are awake, you are still going to feel that this is different from what is happening in the human being.

I don't want to talk in circles, but it seems to me that if you have those four terms ("soul," "mind," "consciousness," "thought") and you want to pin them on a board—they are all so interrelated that taking one out will undeniably affect all the others. In other words, we are constructing a matrix of what the mind seems to be, thanks to the labels we are attaching to it. And we have no way to know what a *nonmind* is.

LAURENT. The dialogue we can have together—or the dialogue I can have with myself—is also a way for me to know what "I" mean. The internal process of dialoguing with oneself allows one to get crisper ideas, to change, to weigh them, and to choose between them. I am not so sure to *know* what I "mean."

SUE. But what you describe is still an internalization of what you have done with another individual.

LAURENT. Correct.

SUE. Other individuals have taught you to self-dialogue in that way, even, in your case, in a very sophisticated way. There are human cultures in which that, I think, almost never takes place.

LAURENT. The project behind the dialogues we are having relates to the philosophical and theoretical breakthrough that is tied to dialogue in classical Greece.

SUE. Yes.

LAURENT. Within the early history of "philosophy"—that is, a mental activity beginning at the latest in Greece, in the sixth century BCE—we find a significant innovation that is traditionally ascribed to Socrates. Socrates is this man having dialogues with friends, peers, enemies, and regular people he meets on the streets or at the "gym." This particular practice of philosophical dialogue outside the settings devoted to teaching leads to no writing by Socrates, but it later becomes a written genre. And maybe Socrates was not the first one to introduce dialogues in philosophy, but let us put this aside. Of the dialogues featuring Socrates, we mainly have Plato's and a few others written by Xenophon, who was also a disciple of Socrates. There were other authors—whether they had been followers of Socrates or not—who were writing dialogues, either with Socrates as a character or with just other people discussing philosophy together. It is believed that Aristotle composed dialogues of the same type, even though we largely lost them. In the fifth century BCE, the discipline and practice of philosophy was already firmly established in Greek cities, islands, and colonies. The further adoption of dialogue as a discursive genre (instead of a recourse to aphorisms or poems, as had been mainly done before) has shaped the entire tradition we still inherit in the humanities, in the sciences, in our own political institutions. Now, *before* dialogue altered the *praxis* of philosophy, dialogue has been equally crucial to the advent of tragedy and comedy. And we also know that, for democracy to arise, one needs people to exchange with each other, and not only have one tyrant—or a group of "talking heads"—telling others what to do, feel, and think.

One of the fundamental ideas that animate our own collaboration is related to the fact that, at one particular point in the history of the Western world—one point that is often seen as a benchmark, a milestone, or a deceptive origin—dialogues changed the way people acted in the political, philosophical, theoretical, and literary realms. A *dia-logue* goes *through* us. It is not only that I am speaking and you are speaking; something is going through us as we dialogue.

SUE. We must change each other. It is like building together a multidimensional object.

LAURENT. Exactly. What I call the intellective space is the kind of ad hoc and transient space where we think together. And, of course, what we think in this supplementary manifold has to be internalized afterward, very rapidly, in terms of milliseconds. But we are internalizing what lies outside of us. The space we are in, through dialogue (and not only through conversation), is a way for us to exchange meanings with one another, within ourselves, and within our selves. Now, there is a certain illusion that goes with philosophical dialogue. It is widely represented in Plato's work and it advances that, by talking and exchanging together, we would completely agree on a definition. In Plato, nonetheless, irony is almost everywhere, so everything could bifurcate all the time. Even if we agree on a definition, the ironical tone could lead the reader or the audience to believe there might be more to it than what has been said. Still, the instrumental use of the dialogue is maintained. A large part of the sciences is devoted to the extreme reduction of "noise" in the signal, or of what would lie outside the "common core" of the meaning. This approach is a preliminary to the use of formalized languages. In a verbal discourse, the less referential words are, the more their meaning will be constructed through other words. Each time we add new words and new meanings to the new representation, we create something that develops in a nonlinear way. There is much more motility and dynamics in terms of semantics with words that are less directly referential. When you "compute" all these words together in a text, the effect is greater. If it involves two or three persons rather than one, this phenomenon is even vaster. It affects all the words we can use together, including consciousness.

SUE. This kind of computation was what formal grammarians were after when they tried to work with Panbanisha. But they wanted repetition to prove their point—and repetition was dissolving meaning . . .

LAURENT. The fact that consciousness, which seems to be a possible name for the state we are both in right now, would sometimes not be treated as such in philosophical elaboration is certainly significant. Even though the words *mind* in English, *mens* in Latin, or *nous* in Greek could have very divergent meanings by themselves and in the texts that consolidate their semantics, they are ubiquitous in philosophy. In contrast, *consciousness* is often absent from texts conceptually addressing *thought, mind,* or *soul.* I was proposing that such an *intermittence* would not be historically contingent but might well reflect a feature of consciousness itself.

SUE. Are you speaking in the way Julian Jaynes is talking about consciousness not being there in earlier societies pre-*Iliad*?[2]

LAURENT. We could mention those claims that have been made several times (with, usually, divergent chronologies), that consciousness would arise not in prehistorical times but in historical times—or not before the Western Renaissance—or that it would not be found in every civilization, etc. Such claims belong to historicism and cultural relativism. That consciousness *as such* would be historically constructed through and through would neatly explain the odd history of its concept. But I am afraid this line of argument is mainly logologic.

I rather wish to interpret this historical intermittence as a consequence, or property, or attribute, of what "consciousness" could be in a transhistorical perspective. The brain of the dreamer is conscious. It is not conscious that it is dreaming, except in lucid dreams. In both Descartes and Zhuangzi, one finds conjectures about the illusion of self-consciousness: as I am speaking to you, I might, in fact, be dreaming that I am conscious and that I am not dreaming. Those ideas, which also pervade science fiction, from Philip K. Dick to *The Matrix,* equally hint at the evanescent limits of consciousness.

SUE. The way the word *consciousness* is mostly used today is very synonymous with *attention* or *awareness.* I may be conscious of everything in this house in a certain sense, if I focus my attention on the phone, on the backpack, or whatever. I may, but . . . At this moment, I have an awareness, including an awareness of my own activities and intentions, of your activities and intentions. If I were not aware of your activities and intentions, I couldn't behave in a normal, social way. I don't have to look at the telephone and care about its intentions. But I have to make some kind of assessment about the intentions of every living being I am around. If it is a nonliving being, it does not matter. If it is an animal, I can't inquire as to what its intentions are, because it couldn't answer my queries. The animal might be running, or slowly walking, toward me, and I can't determine why it is doing this from its own perspective. I have to make a judgment about what is going to happen next. When facing another individual such as yourself, I *can* inquire as to your intent, and, in so doing, I am making the assumption that you are conscious *of* your intent—therefore that you have conscious control *over* your intent. Not only are you conscious, you are conscious of your intention and you have enough of a free will to shift that intention. Oftentimes similar concepts will not be attributed to a nonhuman being, which is one of the basic reasons that people are, for example, fearful of nonhuman beings: because they can't negotiate a region of safety with a being that doesn't have intent or control over its intent, and when there is no communication about that intent.

LAURENT. According to you, we would have something that we could call consciousness, with some level of awareness, some level of control, control of intent, etc. And maybe something that would be a reflexive loop—the possibility of being aware of being aware.

SUE. We normally expect humans to be aware of being aware, once they reach linguistic competency. If they don't behave as if they are aware of being aware, they will be differentially treated. Here, I can use a personal example. When I swim at the Y, there are often groups of adults that live in a special home. They have different mental capabilities and you can see that some of them are trying to communicate with, say, a "normal" person that is with them: you can't understand the words, but they are exchanging glances, they are nodding their head—there is some dialogue going on. Others might also be trying to say or convey something, though it is totally unintelligible. But there is a consciousness on the part of the person who is trying to talk and on the part of the person who is trying to listen. There is an awareness that is intentionally shared, with mutual intent. Through dialogue, mutual intentionality is striven for—even when you cannot understand the speech. Some of the individuals I am referring to may be able to utter sounds, but not to establish a visual glance in an intentional exchange, in which case there is no dialogue going back and forth at all. If they move toward an object in the pool, they just move toward it. Those people are awake, their eyes are open, they can make noise, they are human in shape. Still, it appears that they are not conscious of being conscious. Therefore, you can't discuss their intentions with them. All you can do is *modulate* their intentions, *modulate* where they go, *modulate* what they do by being there with them, thus having some control over them. I would suspect that it is about the same thing with a little puppy: you tell the puppy to sit down, you teach it to be quiet at certain times. You, the human being with the conscious intent, are still modulating the dog even as it grows up and becomes sophisticated. Some human beings with retardation can come to fit in normal society and do things that appear normal. You can cause them to pay more attention to you for what you are going to approve or disapprove of. But this is not the same as your establishing a joint intention through the modulation of a consciousness of being conscious.

LAURENT. I am trying to combine three terms you used in your first qualification of consciousness: *control, exchange,* and, say, *loop,* or some *recursion.* The incapacitated humans you mention are aware of their surroundings. But their control is either insufficient or wholly lacking, so we might help them through, when we are directing their attention to this or that object or feeling. We might "grant" them some control, but indeed control is largely, or totally, outsourced. Plus, there is no guarantee that there would be a reflexive layer (a loop) in those externally monitored minds, that there would even be the ability of *controlling control,* of being aware of being aware. As a starting point, we could certainly assert that those three mental features (control, exchange, loop) need to be interlinked or integrated to give way to "human consciousness."

In posing this, we are also implicitly taking into account the fact that consciousness is not a permanent state in which all humans would be. We are not saying that consciousness must be the only mode of being aware that humans could have. I am noticing it, because of the huge philosophical reaction against the primacy of consciousness. We are saying today that consciousness is not the only mental state that we know of, as humans. We do not even suggest that consciousness is the ne plus ultra of human existence.

SUE. Hmm, *I* would suggest that!

LAURENT. Okay, you would, certainly! But, at any rate, we are reaching this point: there is something that we might wish to call "consciousness" and that we could attempt to investigate and perhaps evaluate against the idea of our lives being controlled by "the" unconscious.

SUE. Good. Let us now take an artificial intelligence program that is trying to talk with a human agent. The software could use phrases that are typically used to discuss a given topic in normal conversation, and you are not really trying to probe anything. It is very difficult to tell if you are talking to a conscious being or a computer, but we are all going to assume that the computer is not a conscious being, correct?

LAURENT. Correct. We might agree that my laptop in front of us, even though it is recording the sound of our voice and maybe doing other tasks in the background, is *not* conscious. This being said, we also know of definitions of consciousness that are so minimal—such as "consciousness arises when a mind processes information"—that a thermostat could be described as being conscious.[3] People could argue that we just broke the concept of consciousness into awareness, controlling, and recursion—and that one could definitely invent a software that would at least be *related* to some, or all, of those three dimensions. This discussion of consciousness is very tricky. It has to be this way. But the fact that sometimes we are *not* conscious might be a way of approaching the issue.

SUE. That is what I tried to say at first.

LAURENT. And that is why I was insisting on the intermittence of the concept.

SUE. We may also need to differentiate *not conscious* from *unconscious.*

LAURENT. One could wish to use the latter term to refer to the built-in intermittence of consciousness, whereas *not conscious* would be used for any organism or machine that lacks even the ability to develop the high-level consciousness you speak about. Yet, I am not completely sure we would easily stick to such distinctions.

SUE. Let us try to go further and admit that a computer could be "conscious." In this case, the only way it could become "unconscious" is if you erase its memory or if you turn it off. One could erase human memory. People could lose their memory: they can go to sleep and wake up; they are still conscious. It doesn't need to change what we think of as basic human consciousness. What changes what the external observers think of as being conscious is when there seems to be no self-reflection on the behavior in the mind of an agent. This doesn't imply there is no modification of the behavior, or no learning. There is the mere execution of the behavior in a successful or unsuccessful way, much like a neural network would operate. So we have a first level of consciousness: a level of awareness that allows behavior to be changed. We reach another level of consciousness when the behavior can be changed by the producer of the behavior.

Let us say you write a very sophisticated artificial intelligence program. Then, you write another AI program whose only job is to modulate what the first AI program is doing. This is almost a non sequitur, because the first AI program only gets better by behaving in some way, then taking feedback and having that feedback put into the program. The program is not really *self*-modulating. All that is really happening is this: the program is doing something and it is being either understood or not understood; and there is some change being made in the program because of the outcome. Even if you use the term in a loose way and consider there is some self-modulation "by proxy," you still cannot put another program above that program to modulate that program, because that program is already "self-modulating." If you were to put this other program, it wouldn't be doing anything different than what is already happening at the level below. And you are suddenly entering the debate surrounding qualia, the homunculus, and the like.

In humans, what is monitoring the *me* that is the experiencer? What is this dichotomy between the experiencer and the thing that is monitoring the experience? Who is actually having the experience? This is not really a question that anyone knows how to address with a computer. If an agent is having an experience and changing it because it has the experience—whether the agent be a computer or a macaque—I guess you could call this *awareness,* although I am not so sure awareness would apply well to a computer. In artificial intelligence, it takes a programmer to come in, look at what is going on, and reflect: What are the limitations of this program? What do I need to do to give this program more capacity?

We apparently have a similar configuration in the human brain, when this part turns on and becomes conscious. I become conscious of the *me* talking to you.

I become conscious of the *me* going to sleep or waking up. Who is the *I* that becomes conscious of the *me* that is doing this? Is it another layer of consciousness layered onto that consciousness that is already there? And if you want to stack other layers onto that one, technically, it is possible for humans. I can think about me, and I can think about me thinking about me. So is the ultimate recursion Chomsky talks about. Chomsky will say: *recursion* is the essence of grammar, it is the essence of human mind. Well, if it is, what is it? It is not just grammar. It is being conscious of being conscious, with different levels of consciousness of being conscious. The first time you are riding a bike, you are just trying to balance it. Once you have learned how to ride the bike, you never try to balance it, and your whole consciousness of riding the bike is completely different than your consciousness of learning how to do it. Your consciousness has shifted someplace else. You are thinking about where you are going. You are not thinking about pedaling. You are conscious of it and you could give feedback. But this is not where your attention is. Your attention is somewhere else, and you feel yourself to be somewhere else, you locate yourself somewhere else. If a Buddhist monk has *n* different states of consciousness, as some neurological studies on meditation would indicate, he is locating himself in *n* different places.[4] Some kind of internal awareness of the self lends us an ability. Like when "I feel sad," or "I feel happy," or "My finger doesn't hurt and the cut doesn't hurt anymore." There has to be some experiential embodied perception of the *mind*—or of the *soul,* or of *thought*—in the same way there has to be of a cut on the finger that is allowing us to even think that we can attach some part of ourselves to some words.

LAURENT. We would propose that consciousness is coming with degrees (awareness being one level). There, clearly, "the" unconscious cannot be seen as a "dark continent" that would control or influence "the" conscious. We mainly deal with a question of *access.*

SUE. I would speak of *levels* of access.

LAURENT. Granted. We now find within the degrees of consciousness the possibility for recursion and self-relation. Even though the self is not identical to consciousness, it seems that we need to elucidate a bit more what recursion and self-relation are, if we want to understand the nature of consciousness beyond the level of awareness.

SUE. In human beings, there is, through the vehicle of language, this recursive ability to think about the self—even without any action whatsoever. There is an ability to be aware of the self, and aware of the feelings of the self. You can even say to a person, "You feel sad, but you shouldn't, and here's why you should feel happy." You can use language and change the emotional perception of someone else's self. This ability is so quintessentially tied to language that we can't imagine it in existence without words. I could ask you to access your past feelings, your autobiographical history, or your current thoughts; you would immediately understand what I said, and you could oblige. In order for one part of the self to assess another part, one needs a conscious self that is aware, and aware of being aware—then boom! Consciousness doubles, and now this other consciousness is assessing that. Then, we have the capacity to say, "Instead of thinking about how I think about things, let me ask Laurent what he thinks about what I think about those things." And this is another level! Then, let me think about this: what if I were just up here in the ceiling watching this conversation thinking about what Sue thought of what Laurent thought about what Sue said to Laurent. Such a removal, in a conscious way, of whatever the attention is being focused on, could go on forever, essentially.

But the normal way of acting for an animal—and oftentimes for a human being as well—is to focus its attention on a goal, and act toward that goal. There is almost something extraordinary in the way human consciousness—

once beyond the basic level of awareness—is achieving its goal. There is something extraordinary that can happen, once you take yourself out of the here and now and recursively reflect upon it. Possibilities become endless. Perhaps you are only doing this a fraction of the time. Perhaps you are unable to hold your attention enough to do it for very long. The fact remains that you can do it. Would a computer do the same? Or can we ever, in the future, program a computer to do it? My answer is no. In the future, we could program a computer that would look as if it were doing those things, but this could not hide the evidence that the programmers themselves would have created that level of computing. The computer will not achieve this on its own, no matter how much you let it run.

LAURENT. The proponents of what one calls "strong artificial intelligence"—who routinely affirm "Oh yes, we can write such a program, maybe not today, but in a matter of years"—preemptively erase and empty out a lot of the features that we would believe to be necessary to human consciousness. The usual approach of strong AI is based on a very reduced version of what human or animal consciousness could be. Interestingly, even Shannon, the pioneer of information theory, once wrote that "Turing's definition of thinking" "does not reflect our usual intuitive concept of thinking."[5] Shannon was critical of Turing's saying that the main difference between a human brain and a computer was an issue of "storage capacity," in which Turing was even more of an extremist than someone like von Neumann. At the roots of artificial intelligence, one finds this ludicrous assertion about storage size being the determining factor for all "diversity of behavior"—an assertion that is coupled with both a theoretical and a practical reduction of what we are and what we think. One also finds, in Shannon and others, the intuition of a different "concept of thinking." What could be called human consciousness does not proceed from the mere addition of storage capacity nor from parallel structures with simple feedback. Mastering recursion allows us to reach a new level of consciousness.

SUE. We spent many years looking at counting in chimpanzees, and doing all sorts of studies about what counting is, to see if there is one-to-one correspondence, or seriation, etc. I always stated that once the concept of counting is understood, a mind could just extend it. In order to really understand the concept of counting, you have to grasp that you are enumerating things and that enumeration is a process. The process itself is not delimited by the numbers. This is where the concept of infinity comes in. The chimpanzees learning to count should get to a point where you could introduce a new number, and, immediately, there would be no errors. For Sherman, who was the chimpanzee I worked with for this task, it took getting to about twenty or twenty-one. It takes that level of counting to *demonstrate* the principle, if one can't *explain* it. Finally, Sherman "got it": any new number that was introduced to him was just one more than the previous one. Numbers were no longer separate entities. Sherman could go on counting without having to be taught numbers one by one. If you think, then, of what counting is, at its basic level, it is a process that has one more recursion. And what is the species that counts? Despite differences, the process of recursion in counting and the process of recursion in grammar are linked to this mental ability to fly free from the whole realm of conditioning, learning history, whatever. It takes you out of that realm.

Duane Rumbaugh started looking at this in his work. I don't think he thinks of it this way, but, in the transfer index, he would give monkeys or apes a problem of this type: you see these two shapes, a white one and a red one, and you have to pick the white one, or you see a book and an apple and you have to pick the book, etc. You learn that with each problem having two possible solutions, only one is right. Then you just shift it: the solution that was right is now wrong, and vice versa. If you conduct this experiment with monkeys, first of all, it takes them a very long time to learn that there is always one right solution. Then, once you make a shift and hold to it, it takes monkeys another very long time to understand and learn it. Those are two different tasks for them. The monkeys can learn to switch, but they keep trying to do what they did before. Then you give them what has been called a learning set: more and more problems that are similar to the problems you just gave them. Then, they become more and more able to do it, and to make the shift faster. But humans or apes with language essentially learn quickly to make the shift: all they need is one experience. They have one experience where they see that the conditions of the trial have changed, and they never make the mistake again. All of this training, all of this conditioning history, can be made irrelevant in a second. You just give me a new piece of information, and I don't need to repeat what I just did all that time. Now, I can behave in a completely different manner. I am assessing that there are rules or parameters by which the experimenter is operating. (Language, by the way, is basically *rules,* or symbolized behavior, as opposed to *conditioning.*) My job is to figure out what his rules or parameters are. As soon as he gives me a clue, it doesn't matter if I have done this task a thousand times before, I can do something else, because I am monitoring me, I am monitoring what the experimenter does and I have learned his rules. I am monitoring the relationship between what the experimenter does and what I have done across trials and across time in sequence. Any organism that has such a capability is freed from its history: it still can access it, but it is virtually freed from it. A lot of studies with retarded kids show that until they get a modicum of language, they can't do it. As for apes, some learn it very quickly, and others learn it very slowly. Most of the data presented on apes by other researchers does not consider if the tested individuals even had a modicum of language comprehension. But apes can definitely come to do it, and apes that have language come to do it very quickly. It does not seem that other species do it.

LAURENT. The counting experiments you were describing could lead us to examine the following hypothesis. Maybe the possibility of reaching the level of recursion exists in many animal species, including nonhuman ones, but the ability to consolidate the bifurcation and to turn this particular consolidation into a stronger reflexive structure (operating beside counting for instance) is being tremendously supported, or even created, by language. In short, does bifurcated consciousness arise because of language?

SUE. The answer is *yes.* Consciousness is arising because of dialogue, and because of the function of dialogue itself, which exceeds language. Language can exist in some form without dialogue. Think of those stroke patients who don't know their legs are theirs, or of the "split brain" where the right side doesn't know what the left side is doing. The left side of the body can do something that the right side is completely unaware of. The right side can see this, and it can confabulate an excuse for why the left side did it. This split-brain person is clearly "conscious," he still has language, but he no longer has the kind of internal dialogue we were speaking about. It is dialogue itself that is creating something that can't exist without it. At first, dialogue is external, then it becomes internal, and as it does so, parts of other people become parts of "me."

LAURENT. Is dialogue consolidating something that was there (dormant, maybe), or is it creating it?

SUE. Through dialogue, we create a new being, and/or a new beingness. You know the view of Andy Lock, that, through language, parents and infants re-create the culture anew at every single generation.[6] His view is very much that it is an actual creation and he gives really interesting examples, like when he studies infants in a bassinet. Some mothers are reaching out and saying, "Oh you're so cute, oh you want to come up to me *[claps],* oh come see mommy," and then they just pick up the baby. Other mothers act the same, except that they are waiting for the baby to make eye contact, or the baby to wave its arms, to wave its feet or some gesture they interpret as the baby having shown some kind of mutual intent and mutual agreement. *Only then* will they pick the baby up. Thus, the baby is physiologically able to straighten itself, so when the mother picks the baby up, the baby *knows* it is going to be picked up and it is ready to be picked up. In studies of mothers in videotape, Lock found a huge difference in how mothers pick their babies up, in how they signal to their babies they are going to be picked up, and in whether they create (or not) the space for the expectancy that the babies will signal back. The creation of a space for the baby's self-expression, or the lack thereof, has, I believe, lifelong consequences.

Let me take this to the next step. Here is a chimpanzee mother, her baby is sitting beside her, and she hears a noise in the forest. Is she going to say "Come here," is she going to gesture to the baby, or is she going to rush toward the noise? If the baby is clinging to her, the mother chimpanzee is just going to rush toward the noise and the baby is going to cling very tightly. When the baby is reacting this way, the mother is not creating any kind of dialogue with her baby. Let us imagine that I take a bonobo baby and put it down near me, or that I lay it on a blanket, so that it does not need to cling to me, what is happening? Panbanisha would do this with Nyota, but if Matata is living in a tree, she can't: she has got to keep her knees up, or that baby, for its own survival, needs to hold on to her. If there is enough protection and the mother can be on the ground and put her baby on the ground, then, when she hears something, she can signal to the baby. If she signals to the baby and waits for it to signal back, a dialogue has been created. An intentional state of understanding in this baby will not appear if there is no dialogue. It may eventually be implemented when the baby is older. But if you are aware of those facts and you intentionally facilitate dialogues from birth in the way we did it with Panbanisha and Kanzi, then yes, you are *creating* this ability in a substrate that has the capacity for it. If you are creating this very early on, you can branch it off in all different kinds of directions. Would we able to do the same with a baboon baby? Here again, we don't know.[7]

LAURENT. Thus, we need some capacity in the organism. However, the capacity is not only "expressed" or "translated."

SUE. It is developed by culturally, dialogically interfacing with the other.

LAURENT. There seems that all civilizations of *Homo sapiens,* possessing similar capacities, also implement at least *some form* of dialogue, which would lead to the huge commonality of "human consciousness" we can all perceive. But should this imply that even in historical times, the *kinds* of dialogues we have among each other would contribute to the *kinds* of consciousness we could develop?

SUE. Yes. I can give you a very concrete example of a human society that I was fortunate enough to be enmeshed in when I went to Wamba, in the Congo region. In that society, people go out to the other members of the group and they just start talking, as if everyone there were listening—but they're not: people are walking past and doing their own things. Individuals may have very strong feelings about which person they would like to have a dialogue with. Or they may have a topic they intend to discuss. But the best I could tell, the kind of dialogue that is typical of the society I grew up in didn't exist in that culture. Dialogues about where we should go to find certain kinds of plants, dialogues about how to make palm oil, dialogues about where to hunt, even sometimes dialogues about Mobutu and other recent events—all those were extant forms of dialogue, but there was no "abstract dialogue." It seemed that this kind wasn't there.

I was trying to understand what the people who had followed the bonobos for years and had been paid as trackers knew about the apes. I would ask questions from time to time, and it became clear that if I asked the question where there were two or three people, each one would just immediately respond, thinking that they knew the right answer and all they had to do was to give it to me. At times, it also happened that the people I was asking were not in mutual agreement: they just looked at each other, amazed. They couldn't even believe they didn't have the same answer. They had never thought that they didn't have the same answer: they never had a discussion about the topic. When I pointed out that they sometimes had different answers, the fact was just extremely puzzling to them. They would start to dialogue together, to try to figure out why they had those different answers. They were people who had followed the bonobos for twenty years, who thought they knew what the bonobos were doing—but this kind of dialogue, this kind of *aboutness,* about the individuals they were tracking, was a type of dialogue that had never occurred to them before. Once I began asking those questions leading to different responses, the dialogues were started, and they would last across time.

In the first part of my stay in Wamba, the forest was this huge thing to me. I was completely lost, I could not find my way, I did not know what was going on: every minute of my life was dependent on the trackers and I trusted them immensely. After a while, the feeling of total disorientation weakened. I began to realize I had already visited this or that place. I noticed several times that there were nests in some areas of the forest, with no bonobos. I came to ask the trackers: "Do the bonobos go back and make nests where they made them before?" Some said no, some said yes. And they talked and talked and talked, and they finally agreed that yes, it was true, the bonobos did go back and make nests where they had made them before. Then I immediately began to wonder: "Do they get in the same nest and do they make that little bed again? If this were a home base, how many places do the bonobos have where they make their nets? Do they always go back to the same nests eventually, or are they actually making nests in new places?" The men started talking. I couldn't understand them then, and I thought they had ignored my questions. They were eight people, and they talked about the issue all night long. They came to me in the morning and announced there were fifty-three places in the forest where the bonobos made nests. Once that dialogue got started among them, they started asking all different kinds of questions about the bonobos, including about intentionality. Once you have the breakthrough that is such a dialogue, it adds something that forces individuals to come to characterize the world in a converging manner. Or to attempt to do this. This is what cultures do all the time. They force us into ways of cooperating and coordinating, many of which we are left unconscious. But when we have a conscious dialogue about those things, we can lift them to a level that we would be unable to access if we didn't have a conscious dialogue. This might be what happened with the ancient Greeks.

LAURENT. The moment one is in dialogue about something, one can be in a dialogue about everything, including about what is a dialogue. If we have a dialogue about dialogue, then everything bifurcates all the time and we can reflect on everything in a very different way. There might be a tension between this sort of conscious dialogue and the unconscious prescriptions of a given society. A society may also decide that some topics are off limits. All this actually happened in Athens as well, when Socrates was sentenced to death on charges of corrupting youth and disrespecting the gods of the city. But the moment the noetic device is being appropriated, it is also the case that you can have a dialogue with people you don't know, and you can have dialogues with yourself. There is no predetermined limit in terms of the number of people you can dialogue with in the course of your life. And, thanks to writing especially, you may have a dialogue with the thinkers of the past. You can multiply, and this also one of the things writers do when they create characters. You can multiply dialogues, *n* times.

SUE. Yes.

LAURENT. As I briefly said before, the breakthrough of Greek theater is coming when, instead of having a chorus (with maybe a leader or soloist), one, then two, then three actors are added to the group: they now interact and exchange. Each actor plays several characters. The difference between the tragedies and comedies of the sixth century that we don't know so much directly and the later ones that we have precisely lies in this sudden expansion of performed dialogues. If there is such a multiplicity onstage, there is a structural need for discordance, dissonance, as well as for unison. There, dialogue is a way of thinking.

SUE. Presumably, dialogues existed socially before actors were added onstage. The plays were just *depicting* dialogues, weren't they?

LAURENT. Presumably, yes, the social form of dialogue was extant before the tragedies and comedies I am speaking about. But theater did more than *depict* a collective and noetic practice; it made it more readily available to the audience; it fostered it. I may need to stress the fact that stage performances were the only event in Athens that could be attended by all, including women and slaves.

SUE. I see.

LAURENT. Now, I wonder about the kinds of dialogues we can have *with* the bonobos you raised, and those they could have among themselves. Would you say that they participate de facto in a dialogue, through the dialogic culture that you cocreated with them? To qualify my question a bit more, I would like to make one observation. During my first visit to the lab in 2010, I was not proficient at all with the lexigrams. This meant that I could speak in English, but if the bonobos were talking back to me by pointing at symbols on their paper keyboards, they had to be very patient and give me enough time to see which words they were using. This was a clear limitation to any verbal interaction we could have together. I remember that, in 2010, after I asked Panbanisha a third time to repeat the three lexigrams she pointed at on her paper keyboard—because I just could not map out what she was saying—she folded the three panels together, looked at me in the eyes, and, from that point on, communicated with me only through indexical gestures and glances. Beyond this issue of language competency, I also felt that neither Kanzi nor Panbanisha was a priori interested in exchanging with me. They were studying me, and it seemed there were preconditions for any actual dialogue to emerge, whereas in many (though not all) human cultures, some form of verbal exchange is rather a prerequisite for social interaction. Usually, in a first informal encounter, we stay at the level of small talk, so we are rather engaging in low-level dialogues, assuredly not the one you and I are having now. But the rudiments of a dialogue are there. With the bonobos, I felt that the more we knew each other, by spending time together, the more dialogues we would have. The possibility of having a dialogue had to be earned, in a sense.

SUE. I would say more precisely that the possibility had to be co-constructed. It *feels* to be earned, because it does not come as naturally as in a dialogue with other humans (which was historically co-constructed or "earned" long before you were conscious of so doing). You also have to come to treat bonobo expressions or gestures with the same validity you would assign to human ones—and this, too, is "earned." When Bill Fields, Pär Segerdahl, or Itai Roffman speak about their first impressions with the bonobos, they all say exactly, word for word, what you said: this has to be earned. Those persons like you have a deep comprehension of the philosophical literature and of the primate literature: they didn't walk in expecting a "normal human situation." They are open, you know, which a lot of people aren't completely. So, yes, you need to have those preconditions, and if they are in place, then people generally will have an experience similar to yours. However, those preconditions are not in place with, say, a reporter who comes to the lab. They are not necessarily in place with a lot of employees who come to work at the lab. To get those conditions in place with a reporter or an employee might take years of reading and watching videos. In their lives, these bonobos have been mainly exposed to people who do not come with those thoughts and ideas and with the abilities to open up, to look and listen. So, they are going to watch if you are one of the people who really can pay attention to the subtleties of their communication. If you are not one of those, then you are simply not worth their time because you will *not* get it, no matter how much *they* try. A lot of people don't make it very far. Even if they stay and work with them, and even if they become good friends with them. Many people can't make it where you have made it in two or three days. They can't go that far. I just think of it as a perceptual blockage. Some people can see in color while some people can see in black and white. If you can only see in black and white, it is not really going to do any good to lecture you about color. But if you want to know if a person can see in color, you can do subtle things to determine if color is attracting their attention. In the same way, some people may just have a block about what nonhumans are able to do, or about any kind of languaged being that is not using "speech" the way they do: those people don't reach beyond what they expect, and if you don't meet what they expect, you can't do it, and "this is not language." So yes, the apes were testing you, like you were testing them.

LAURENT. I felt my own perceptual block at the beginning of my first visit. It didn't take me long, but I maybe had to wait for one day, or one day and a half, before getting that when the apes were nodding with their head, they were saying *yes* as a direct answer to a question. At first, I guess I was physically able to see it. I just could not understand it.

SUE. Their necks are shorter, so when they nod, it is not so perceptually obvious to someone accustomed to the skinny human neck.

LAURENT. Yes . . . it was not obvious to me . . . Furthermore, by coming from the outside, one would not automatically suppose that this gesture is semantic in the same way it is in our human groups. This leads us back to your own assessment about the consciousness of those bonobos. If we believe that dialogue is creating consciousness, I guess we would agree that consciousness is not something that is identical all the time, and that, even at the individual level, the kinds of dialogues that we went and go through do shape our consciousness. At any rate, if the languaged bonobos have the singular relationship to dialogue we have just described, this should imply that their modes of consciousness depend on the kinds of particular interactions they have.

SUE. Yes.

LAURENT. How would you describe it?

SUE. How would I describe the process of how their consciousness is being shaped, or how would I describe the process of their dialogues?

LAURENT. I guess both are interrelated, but since we spoke about the process of dialogues, we might concentrate on the way it impacts ape consciousness.

SUE. It is only in the extension of possibilities that a dialogue can take place. If you didn't extend certain competencies to me and if I didn't extend certain competencies to you, this dialogue would be short-circuited. Such is the cause of nearly all human problems, that is, the failure to extend sufficient competencies to the "other" for any dialogue to commence. The answer to your question is not just what the apes do; it is what Sue was able to understand or discern that they were doing, and then what Sue was able to communicate to other humans. I can say that, across the four decades that I have worked with this family, the biggest change has probably been in me. As I make changes in me, through what I am doing, I can create more complex dialogues with the bonobos. They begin to try to communicate more complex information to me. But we are always surrounded by a cadre of people who can't participate, who don't see and don't believe. Even people who are taking care of the apes may be unable to participate. The kind of dialogue that can occur is shaped by me; I later report it to you. All I can say is that it goes beyond any ability I have to describe it.[8] I know that the limitation of the captive environment is not allowing me to see the whole range. I have seen bonobos in the forest. When I was around a group of bonobos in the forest, it was overwhelmingly clear that they had a deep understanding of what was going on one or two miles away. In the wild, groups of bonobos simply know what happens in the forest: they are communicating about it at a very high level. The bonobos I was around in Africa were watching me constantly, figuring out what I was about to do. If they don't like what you are going to do, they can have complete control over it: they can lose you in a minute, they can go to a place and wait for you to follow them and when you get there, you find there are pythons in the tree right over you or a hornets' nest. The bonobos can easily maneuver you into danger; they can easily maneuver you out of danger. They are masters of the forest, and you are not—they are intellectual masters of the forest. They aren't out there just calling and finding food. They are doing all kinds of very complicated things.

There is a sharpness, there is an edge to their perceptual abilities that gets dulled in captivity, because, in captivity, it is so much of the same thing, so much of the routine every single day. It was wonderful to be able to work so much in a forest, as we did in Atlanta, but I would like to work in a forest where even as adults the bonobos would not legally need to be on a leash and the forest would be completely open.

LAURENT. In captivity, human beings go through a dire restriction of what they can do. There are those widespread anecdotes about prisoners who, in the harsh conditions of their cell, survive mentally through the poems they know and recite to themselves, or through the books they are allowed to read. For most prisoners and in most prisons all over the world, there is very little survival of the mind. Consciousness is being reduced.

SUE. This is why you don't want to have isolation. Most of the people who stay in isolation for many years acquire very atypical behavior patterns.

LAURENT. Assuredly.

SUE. With regard to the dialogues I could have in the captive situation in Iowa, which is where Kanzi and Panbanisha actually became adults, their ability to engage in dialogues became increasingly sophisticated. It seemed that when they moved from Atlanta to Des Moines, they saw that the whole world was different, that people treated them differently, and that the things they had come to expect—like always going out in the forest—would not be there. It was no longer the wonderful and artificial world that I had been able to put into effect in Atlanta. I wasn't in charge of the world in Des Moines and I had to cope with whatever was tossed at me. When Kanzi and Panbanisha understood this, they began to reevaluate what the world was like, and what were the problems I was really facing. At that moment, Panbanisha became tremendously interested in documentaries that were dealing with human–ape relationships. Kanzi had always been interested in the NHK documentaries about himself, and he would practice for those, but, in the past, Panbanisha had never shown any strong investment. She would mainly let Kanzi do his thing. In Des Moines, Liz and I were watching those films in the lab, because we didn't have much time: we were watching the movies we wished to study, while taking care of the apes. This is how Panbanisha started watching videos with us, and she became very serious about it. Until that point, she had preferred the daily soap operas . . .

Later, when people like Anderson Cooper would come to film, Panbanisha became very concerned about what she and the other bonobos would do and look like in front of the crew. We could then ask her questions like "What should we do? How should we plan this? How should we plan that?" She would help in every way she could. The first inclination I had about her capability was when we were getting ready for Paul McCartney to come to the lab—it was near the end of our years in Atlanta. That day, we are cleaning everything up, and we are talking about what we, Kanzi, and Panbanisha are going to do—we always have to plan ahead. Panbanisha is listening to everything. Then we think we are all ready and she says "TRASH TRASH," pointing to a place outside the enclosure where we had left a bag of trash. When Paul McCartney comes with his people, maybe twenty persons, there is no focus at all: some are singing, some are talking, some are laughing, some are walking around. Panbanisha tells everybody to "SIT" down, "BE QUIET," and "EAT PEANUTS." And the visitors sit down in front of the cage, they all quiet down, while starting to eat peanuts! Panbanisha organized all this: she knew that the way to calm everybody down was to ask them to have a sit and to eat food! She knew she was a focus of attention and she began to employ this situation in order to appropriately structure a group of important guests *through* language. A bonobo can get more "high level" than many humans in the same setting. Panbanisha could take competent control over a human group to the condition of humans extending this possibility to her.[9] Throughout the whole session with Paul McCartney, Panbanisha just guided this group by her comments about what would be the next appropriate activity for us to do within this kind of general interchange and dialogue we were all having together. It took me far too long to recognize such a capability in her. I think in retrospect that I did not use her skills nearly as efficiently as I could have and as I now wish I had done. My assessment is that I have been constantly underestimating the bonobos. I am trying not to underestimate them, but this is what I have done.

LAURENT. At the same time, your critics would say that you always overestimated the bonobos.

SUE. People would say that, yes. But they're just looking through their own, self-limiting, foggy lens of human understanding. I am looking through the lens of being in the actual physical space of the bonobos—just like you and I are in the same physical space right now with no wire between us—and what I am trying to have with you or with them is a dialogue that is working, a dialogue that is successful. I make sure that we are getting it, that we are coordinated, that we are moving forward. In order to achieve this goal, I need to have accurate estimation. When I am a little off with the bonobos, and when the dialogue is not working, it is always because I have underestimated rather than overestimated them. When I make a correction, I have to correct by giving them more insight and capability than I had previously given them. This might be a function of the dual fact that they are bonobos and I am human, and that I have a structural tendency of not grasping what they can do. I rather think this is a property of interindividual behavior and dialogue. In an interaction with others, one is always cocreating minds, consciousness, some synchronous behaviors. If you want a level to be lifted up, the only way is, as I said, to extend to the other beings with whom you are interacting more and more understanding and competency. This is similar to what you say at the end of your last book about the "soul" coming out beyond yourself.[10] Much of the time in human interrelationships, we are not *extending* consciously. I am trying so much more now to do the same with human beings, to do it *intentionally*. Viewing myself not as a responder but as a cocreator, which is what dialogue does for people. It gives you this opportunity.

LAURENT. Precisely, it gives the opportunity, it doesn't automatically give you the outcome. Through dialogue, we have shared a space, we have constructed things together: at the moment we were thinking together, we had ideas that could vanish afterwards, though they might leave an *imprint*. When a dialogue is ending or is being interrupted, you and I take things back, and we both lose some of what we did not *have* before. We also gain new insights. Then, the *extension* of dialogue is turning the latter into something more than a heuristic device. "Consciousness," if we want to use this term, exists in dialogue. The genre and quality of dialogue, then, is shaping consciousness.

SUE. Yes . . . Here, and given the current situation the bonobos and I are facing, we may think of the American legal system, whose structural process prevents any actual or meaningful dialogue from taking place. I understand why the system must so function. There remains that the lack of dialogue in court can at times be a very destructive state of affairs. Dialogues must be pruned, but they must be pruned like bonsai trees—with an eye toward an end product that has been shaped by a desire for beauty, justice, truth, and equality. The "rule of law" is no pruning of this kind.

LAURENT. This is tied to the old distinction between *law* and *justice* . . .

SUE. You come into this world as a human, and you learn how to interact with other humans. You often feel that

> other humans are not doing what you would have them do, or you notice that they are stopping you from what you would do. You blame them for this situation. The whole culture encourages this: it has rules and expectancies and it will make judgments. When I am dealing with another species, I don't really have this option. I cannot blame the apes for what is going wrong. When I was trying to work with Sherman and Austin, or even Kanzi, and get them to learn symbols, I couldn't say, "Kanzi, you haven't paid attention today, you are not listening to your lesson," or "Sherman and Austin, you are not hungry enough. I am going to take away more food if you do not work harder." I couldn't really do that! Even if I did, this wouldn't have gotten me anywhere. I had to reconsider. Whatever I wanted to bring about, I had to think: If they are not getting it, how could I change my behavior that would enable this message to get across to them? The only thing I have to change is me. After realizing this, I decided to try to change myself, instead of telling other humans "You should do what I tell you, and you should do something else." In this respect, I can't say that I completely succeeded in life!
>
> *[We both laugh.]*
>
> But I also continue to improve my capacities through experiencing failures. I think that when I don't succeed, it is because no dialogue is permitted. In legal cases and contexts prohibiting dialogues, one is just wherever one is. People often "stop talking" when they feel they can no longer *control* the dialogue. Then, they often accuse the other party of being "deceitful" or "manipulative." But only dialogue has the potential to change things.

We both laughed wholeheartedly when Sue remarked that she had not always been successful in "changing things" through dialogues. Alas, how true this was! We were four years after our trip to Walmart and the photo shoot of *Green Lantern's Quest for Fire,* and the situation was now completely different. Both Panbanisha and Matata had passed away. The remaining bonobos were staying in the same building as before, but with their symbolic environment, human family, and friends being routinely removed from their existence. In 2014 and 2015, I spent many days working with Sue and her lawyer Bill Zifchak to have her and the scientists of Bonobo Hope regain some authority over the life conditions of the bonobos, and to ensure the latter were not deprived of their own language and culture. Even though we tried hard, there was no dialogue to be had. We were now resigned to enduring the slow and strange procedures of American federal "justice," waiting for the trial hearings scheduled for May 2015.

On the Flavors of Consciousness

In April 2015, Sue spent a week at our house in Ithaca. She came to my research seminar on "poetry and mind," and we all spoke of Constantine Cavafy and Émile Benveniste. And of the mirror test. And, mind you, of apes. Sue met with Joseph, who would later transcribe our conversations. We had a half day of snow. We went to the so-called beautiful Ithaca Commons and stayed an unreasonable amount of time at the mineral store. At the shop, I tried to survive the conversation with the unavoidable "local characters." Sue bought several rocks. We went back home and took our seats in my red living room. We knew that, in its turn, our preceding dialogue had bifurcated, focusing on dialogue rather than consciousness.

SUE. While we are having this dialogue, I want you to hold this rock for a little while and see if you become conscious of anything, or if you feel there is any consciousness *about* the rock or *from* the rock? Unless it's just a rock . . .

Let me say one thing, by way of introduction. As a psychologist, I was first taught I should not talk about consciousness, because it was impossible to study. This was the pure "Skinnerian" view. The point was to focus on what we could see and on what we could control. Behavior was the focus, but mainly the behavior of "the other": the behavior of "the animal" or of other humans, as in B. F. Skinner's *Walden II,* that is still a great, great book, in my opinion.[1] Then, in the eighties, everybody in the field began to say, "Oh, we now need to talk about consciousness! Let's come up with ways of talking about consciousness!" Then came the *hard* and the *easy* problems of consciousness: everybody could make some progress on the easy problem, but not so much on the hard problem. Consciousness became interfaced with the topic of perception. The question became, How does one *explain* consciousness? Or, What is the physical basis of consciousness?

LAURENT. We may want to address some of those problems today, but in an oblique fashion, for there is little guarantee that the conceptual framing of consciousness that currently governs the conversation in analytic philosophy, cognitive science, and artificial intelligence would be exactly what we wish to discuss.

SUE. Yes.

LAURENT. In our previous dialogue, we decided to agree on some attributes of what we could call *consciousness.* We considered that consciousness was not one block: it is constantly bordered by what it is not, and it also varies among individuals, social groups, and species. The concept of consciousness does not have to be reduced, as a matter of principle, to the category of awareness. We also spoke of "being conscious of being conscious." There, *consciousness* became a capability dealing with control, with change and exchange (of behaviors or ideas), and, necessarily, with recursive loops. This is how our previous exploration led to an examination of the structures and functions of *dialogues.* Quite recursively, we proposed that some kinds of dialogues—like the ones we have—could modify and even further consciousness.

SUE. I believe this a very good summary of what we agreed upon last time.

LAURENT. But I am wondering if we should not go backward a little and discuss the role of unconscious thought processes with regard to consciousness. Under the term "unconscious," here, I am thinking of both the psychoanalytic concept and the automated operations of the mind. Then—there is an obvious link—I guess we should ponder the issues of differential consciousness across species by including more animals than just *Pan* and *Homo*. If we do this, I promise to hold the rock all along.

SUE. If you promise to hold the rock, we can certainly return to the relationship between conscious and unconscious. But would you tell me why we should dwell on the unconscious?

LAURENT. Let me put it this way. One could argue that, within modern philosophy and psychology, a growing weight was granted to what *escapes* consciousness.[2] The emphasis on unconscious processes is evident in nineteenth- and twentieth-century psychology. It is methodologically instrumental for very different—and mutually incompatible—research paradigms, like psychoanalysis, behaviorism (with an insistence on conditioning), or cognitive science (since most mental computations are subconscious and automated). What do we do about all this? Have we been, you and I, oblivious to the unconscious—or even "unconscious" of it? My point is not to "dwell" on the topic so much, but to be more straightforward about the reasons why we are setting "unconsciousness" aside.

SUE. The problem for me is this: if we have no agreement on what is the "state of unconsciousness," I have no adequate concept of what we are setting aside. At least in the waking state (and possibly in the dreaming state), there are also mental activities that we are not aware of. Those activities could be described as being *subconscious,* or *unconscious.* However, the existence of such subconscious—or unconscious—thought processes does *not* prevent us from being conscious. There is just a lack of awareness.

LAURENT. There, we make a difference between "the" unconscious and nonconsciousness.

SUE. Certainly. But it really is a difference between "awareness" and "nonawareness." Unconscious thoughts happen without our awareness, but they do not "fail" the fact of consciousness. It is true that they may be influencing our thoughts and behaviors. For instance, if we have been blind and later gained vision, we are trying to learn to perceive a cup, and we have the conscious experience of trying to figure out this is a cup. Normally, people do not have that experience as conscious, except maybe as a baby, but then they have no memory of it, because this process becomes "unconscious." It happens rapidly, yet, without any effort, we are able to perceive a cup as a cup. The process of perception has become automated, or it is offline. But it is influencing our conscious experience of seeing a cup. Freud takes this one step further. He takes it into the emotional experience that you encounter as a child at the hand of adults or people around you—it could be fearfulness, uncertainty, and the like. He claims that those past experiences are going to affect you when you are older in ways that you cannot readily access, because the experience occurred so early that it has become like the perceptual experience of the cup: it is now enmeshed in an emotional network with other experiences that you can't consciously attend to, so you do not fully understand what happens to you.[3] Freud also assumes that we have the same situation in dreams, but with one layer of the onion being peeled back. There are still layers: they are hidden, and even in a dream they do not get peeled back. Thus, he, the psychoanalyst, has to put all of a person's utterances, memories, and dreams together, and he knows how to do this. Even subjectively, even if I am not trying to be a scientist, I am baffled by this claim. For the rest, yes, in perception as much as in dreams, your current conscious processing is at the top of an iceberg of things that happened to you in your life, perceptually and conceptually. Somehow, this immersed part of the iceberg partakes in your momentary conscious experience. If you didn't have this iceberg, you would be like a computer with no memory, and you would have to be constantly learning how to talk, or how to walk. You'd be constantly learning everything, and nothing would be "on automatic." If you have feelings, you are attending to them some of the time, and, at other times, you are not attending to them. When you're younger, you may not have an explanatory framework for your own feelings or percepts, so if you reaccess them later, as you have a new explanatory framework that makes more sense to you, then those feelings or percepts can impact you. Whether or not they can actually impact someone as Freud is describing remains to be seen. I certainly don't have any problem with Freud saying that, in the lives of his individual patients, in his individual society, maybe those things indeed happened as he claims. Do they happen similarly in every culture? Do they happen similarly in every life, with only slight alterations? I don't think so. Every life is different. To sum it up: Beyond what we are calling conscious is this whole other space, which is normally "unconscious"—but if we want to access it, we can. With more or less efforts, we can access some parts of it, maybe even the whole of it. If we have had a lot of education and self-reflection, we can probably access it without Freud or his theory. But having a dialogue sometimes can help.

LAURENT. There are many problems with psychoanalysis, and here is not the place to deal with them. I still felt we had to express that the concept of "the unconscious" is not weakening or relativizing what we may wish to name "consciousness." There is in Freud and his followers an overinvestment in the unconscious that I always found bizarre. First of all, instead of the well-organized inner "topic" of the Ego, the Id, and the Superego—a spatial configuration that, so far, has no correlation with anything we can see happening in a brain—we rather find the moving line of access that marks the border of the conscious and the unconscious. This moving line does not make enough for a "Copernican revolution" to occur. Then, as you just indicated, the alleged universality of the doctrinal elements of psychoanalysis is dubious. Beyond ethnographic differences, the whole issue of "the psychic life of other species" is unexplored by Freud. Moreover, the existence of unconscious thoughts, and the curative virtue of verbalization and dialogues—both being undeniable—actually say very little about the truth or accuracy of the Freudian theory itself. Finally, if I can make myself *conscious of* some "unconscious" object, the examination of consciousness—of the fact of "being-conscious-*of*"—is even more clearly distinct, in both its nature and position, from the "dark continent." And, by the way, "the" unconscious, once said and acknowledged, is becoming, at least partially, conscious.

SUE. Let us look at it this way. There is the patient and there is Freud. There is this big region called the unconscious and there is the executive function of the mind, or the attention. Sometimes the patient accidentally accesses the hidden region and brings it up to attention. More generally, Freud has this great power of vision to access the unconscious to bring it up, and to get the patient to talk about it. This is impressive. But, now, we are going to leave all this out. In this black box down here.

LAURENT. Okay. This is the black box.
[I may have pointed at the laptop. Or at my head. Who knows?]

SUE. We're going to leave this out, all right?

LAURENT. Exactly.

SUE. And we are going to talk about this kind of consciousness, which is often unconscious though it could be accessed by moving back and forth with attentional focus.

LAURENT. Yes, but . . . we may also need to underline the link between unconscious activities and automation.

SUE. If this is so, I want to take out of the box your statement "something unconscious, once it is being said, becomes conscious." This unconscious object may be in your head, or in my head, but we are not aware of it. Here, let me relate the topic of consciousness to waterskiing . . .

LAURENT. Please do.

SUE. At this very moment, we are not aware of waterskiing because we are talking about consciousness in your house—but you can't water-ski unless you are conscious. In waterskiing, everything is happening very fast, and you're having to learn how to get up on those skis and stay up. You don't have the control that you do over a bike, to get off of it or go slower. If you're behind a boat, the boat is determining your pace. You've got to hang on to that rope and stand up at the speed of the boat. What you are learning is to take things from the conscious mind to the unconscious mind. Now, you and I can bring that up by saying it and making it "conscious." But it is not the same "bringing-it-into-consciousness" to think about it in the abstract, as opposed to when you are actually doing it. Or is it?

LAURENT. Honestly, as I was speaking of a becoming conscious, I was alluding to the status of unconscious memories in the psychoanalytic cure. The unconscious and computational elements of my perception as I am looking around this room would not be magically transformed by the fact of expressing them, of explaining them "consciously." They would not even come to the fore, and, *as such,* they would remain hidden to my "conscious" mind. I could be aware of their existence, without becoming aware of them. There are deep, automated, cognitive structures that are far below the threshold of consciousness and normally inaccessible. Waterskiing is too recent as a human activity to be automated as perception is . . . If you allow me to refer to something I know better than waterskiing, I'll move to theater. When you are acting onstage, you may suddenly be remembering that you are acting. This may happen because someone in the audience is coughing, or your partner forgets a line, or you were just distracted, and you stopped being in the role. Before this, you were onstage as a character, not as you playing the character, and perhaps you were conscious *as* a character, or you were in a semiautomated mental state, with your words, gestures, feelings not being consciously determined one by one. In all likelihood, the moment you are thinking you are onstage playing the character in front of an audience, you will forget your line, have a slip of the tongue (though not necessarily a Freudian lapsus . . .), or take one second more than usual to do what you are supposed to do. When you reach a certain level of proficiency through automation, going back suddenly to a conscious and reflexive mode will at least delay your actions, and, at worst, diminish the proficiency you had already reached. And if you are doing the same while waterskiing, I guess you might lose balance and fall.

SUE. Right. I think this is true for driving a car, riding a bike, or being onstage. It is very interesting that we lack the ability to think about the fact that we are acting as we are acting, or we lose the ability to act. But, maybe, if one practiced enough one could think about the ability to act while acting.

LAURENT. Yes. Absolutely.

SUE. I was trying to say that when we are talking about the physical experience of embodiment and trying to do something out there in the world, we must be conscious of a whole array of parameters and facts. We also need desire, intent, and focus. We have to be goal-oriented. But is this similar to when we bring a memory to our consciousness and focus on it? We could become conscious of anything I would bring to mind in this conversation, simply by using a symbolic, verbal command. Through dialogue, we can suddenly have joint consciousness and exchange.

Trying to get up together at the same time in the boat or trying to act onstage would be different *types* of activities, with different relations to mental automation.

I would like to transition our dialogue toward the issue of the conscious mind *as* the symbolic mind. I propose to use the word *consciousness* to refer to the aspect of ourselves that allows us to have control over ourselves, as well as perception of such control—since, as we said before, recursion is necessary. I am adding here that consciousness is a function of our ability to develop a *symbolic layer*. Unless we develop this layer that we then reference down to the flow of the real world, unless we connect meaning between the two layers, we are conscious as little kids walking around. We are not asleep, though we are not conscious of being conscious.

LAURENT. Could we be more specific, Sue? Should we speak of consciousness in the absolute? of symbolic consciousness? of human consciousness? or of reflexive consciousness?

SUE. The consciousness I am speaking about is a type: it has a particular "color," or "flavor," as one speaks of the "flavor" of a quark. Our consciousness colors everything we do, whether it is art, mathematics, or language.[4] To try to conceive of a being that doesn't have the color or flavor of consciousness that you have is something quasi-impossible. Humans tend to put that flavor on what their dogs or cats do. "Animals" can have sophisticated mental aptitudes, but they just don't have them in the way we do, unless they have formed a symbolic layer which can exchange meaning and be self-reflective. To put it in a really simplistic way, this is why some animals can learn how mirrors work without figuring out who they are in the reflection. Which, when you think about it from a human perspective, makes really no sense. If you know how a mirror works, humans reason, then, by every human logic, by every human mathematical calculation, there is no way you could understand that the image is reflecting the outside world around you without grasping that it is yourself in the mirror.

LAURENT. I agree that this is one of the most baffling aspects of the experiment. I remember my surprise the first time I read that a baboon or macaque would understand, after some exposure, that the mirror reflects the environment while being wholly unable to make a correlation between its own body—of which it has proprioception—and the image of the monkey.[5] For us, this is almost impossible to fathom. We might say that our focus is on what consciousness is to "us," as languaged animals. And, to "us," "consciousness" would be typically bifurcated, reflexive, symbolic, and dialogic.

SUE. Yes, and, as we said last time, we achieve this by virtue of dialogue with a caretaker. This is Trevarthen's "primary intersubjectivity."[6] Everything the mother does is affecting the baby, as I said last time. Let me insist on this. Once the baby comes out, when the mother puts it down, when she is smiling and reacting to it, the baby starts to have different experiences. In contrast, the ape baby is so programmed to cling to the mother to survive that, for a significant period of time after birth, it still has to stay with the mother and be one with her. As its brain is developing, the baby ape stays in that "one" state, and its development is normally delayed before it is really able to have its own type of consciousness and act independently from the mother's. As all developing organisms gradually become less plastic, when you add a mental component to what is already there, this is like building a tower: you put a block on top of the tower, and, from that point on, you can't move the block that is underneath or you risk destroying the structure. The earlier you get dialogue and bifurcation of consciousness in, the more you have the ability to build the symbolic layer, that is, the layer allowing one to step into life or step out of life, to refer to life and to reflect on it. You can talk about life while you are in it, but you can't *reflect* on life when you are in it, which is the problem of acting and trying to reflect on the fact that you are acting. Whatever symbolic capacity or grammatical capacity is occurring in life, it is still not in that layer that is . . . "above" life. It is in the layer that is above life that human consciousness is created.

LAURENT. When you speak about different flavors of consciousness, you are de facto positing a consciousness that could be more *this* or less *that.*

SUE. Yes.

LAURENT. We might decide to speak about human or symbolic consciousness as consciousness, for this is the one "we" live through.

SUE. Yes.

LAURENT. But this decision does not obliterate the possibility for other kinds of consciousness to exist. Most of them being nonsymbolic.

SUE. Yes, this does not preclude other kinds of consciousness that are not symbolic.

LAURENT. That are not symbolic *yet,* or that would never have the potential to become such.

SUE. Correct.

LAURENT. Granted. And now I can have my coffee.

SUE. And now we come to dogs, cats, squirrels, and pigeons!

[I come back to the living room with my espresso.]

The other day, as we were discussing together in this room, I spoke of squirrels, and the moment I said this word, the squirrel outside poked its head out of a hole and looked at us. Do you remember this?

LAURENT *[laughing].* I certainly do.

SUE. Was the squirrel conscious we had been talking about it when it poked its head out of the hole? In fact, we don't really know. We could invent all sorts of reasons that it was—or that it wasn't. Then we could design and conduct experiments to test our hypothesis. Psychology, for at least two centuries, dealt with such questions. Scientists mainly used organisms that people don't normally keep in their homes, like pigeons and rats. It is not that they have never seen these organisms before working with them, but they normally don't have an attachment to them. Sometimes psychologists dealt with cats, but cats are notably detached, so well . . . Up until very recently, dogs were pretty much avoided in experimental psychology. At any rate, if I were to make conjectures about the consciousness of a squirrel, everything would need to be proven. If I were to develop the same kind of comments about a dog while discussing with dog lovers and owners, they would tend to say, "Of course, Rover hears me talking about him: he knows I'm paying attention to him, so he sits up next to me when I say his name." Most psychologists would answer, "Rover does know that you said his name, for he's *conditioned* to respond to his name." So, were we to accept the existence of certain mental states in Rover that we wouldn't admit in a squirrel, we'd still argue that Rover is not conscious of his name in the same sense that you might be conscious of your name. Rover is not conscious of reacting to you in order to do something, unless you are going to give him food or reward him or do something that he desires. He is not having a meaningful exchange with you in the way you could have one with another human being. In hearing this, most dog lovers get thrown a little off base. They don't know what to say, because they haven't studied a lot of psychology—and, to them, whether or not Rover's behavior is conditioned is merely beside the point. They do feel that Rover is affective to them, that he likes them, that he has intentions toward them, and that he wants to communicate with them. Then, when they watch animals in the wild, they may feel those other animals are also communicating in a realistic way. There is a great dissonance between what people could feel about animal minds and what the field of psychology is willing to state about the capacity of other species. Let us say that everything that is conditioned goes into the black box of unconsciousness. The additional device that is above unconsciousness and reading, or accessing, the different parts of the box, this is what most animals don't have. If they don't have it, they can't get it. A big piece is missing. As a psychologist, you can't create the piece. What you *can* do is to condition enough of the other little loops and compartments *inside* the box of unconsciousness, giving the impression that "animal" and human consciousness are the same.

LAURENT. We say it again, unconsciousness differs from nonconsciousness.

SUE. Yes. By "unconscious," I don't mean that Rover or the squirrel aren't awake or aware. I just mean there is no access between the two planes. Therefore, Rover can be trained to do things that *look* conscious—but they are not. They have been designed to look like it, which is the same problem that one faces in artificial intelligence or robotics, as I explained another time. Whether we are talking about animal or machine consciousness, we end up in the same impasse. In one case, we are dealing with pieces of silica, and in the other, we are dealing with a living being. We are asking whether they have what we have and think of as consciousness. We wonder if they have the same flavor or color. Maybe we'll call it a *spark* instead of quark . . . Maybe we'll call it lavender. This is the new name of human consciousness: *lavender spark* . . .

In the field of ape language, or in the field of cognitive science where the focus is on computation, the question is now: Could we create this lavender spark we know of in a machine or animal? And if we could, we would think we understood it. But look: we create this lavender spark at every generation in every family, with every child that becomes normal. Human consciousness is being re-created constantly. Is it because of environmental variables, because of culture and dialogue? Or is it happening almost despite them, because there is, in every normal human infant, the essence of humanness? And if there is an essence of humanness, what is it? If we don't even know what it is, then we don't really know what consciousness is, do we? We try to make animals or machines do what humans do, without understanding what it is we are actually doing. With typical human children, it is easy to create: it is "magic." I saw a similar kind of magic—in a much-delayed and long drawn-out form—operating in Lana, Sherman, Austin, Bruno, Lucy, Kanzi, Panbanisha, Teco, Nyota, and Nathan that is in the chimpanzees and bonobos I worked with. Not only did I see consciousness emerge, I acted to enable it, and, in a sense, I created it. To use a rough analogy, I learned how to polish the diamond—without learning how to make the diamond. But I do know a lot about polishing the diamond. Normally, parents just raise their kids and love them. The diamond gets polished and nobody cares about the process. With an ape, especially an ape like Sherman and Austin that has been deprived, or with a child that has been deprived, just trying to be a loving parent will rarely be enough to produce the "lavender spark." There, you have a human being or an ape, and you are aware of its potential—or the latter will not get developed, or it will only partially occur. This potential must be consciously fostered to emerge.

LAURENT. At this point, how can we move away from the caricature debate opposing historicist constructionists and the proponents of the data of "human nature"? What we really wish to abandon is the kind of intellectual dead end that Steven Pinker's *Blank Slate* would exemplify, with the relativistic humanist on one hand and the reductionist scientist downplaying development on the other hand.

SUE. Let us think of a little embryo growing. This little embryo could become Shane, whom I raised as a son, or it could become Teco, whom I raised as a son. I was far better at knowing how to polish the diamond by the time I raised Teco than when I raised Shane. I guess that, had I had to raise Teco *before* raising Shane, I would have failed, and Teco would not have acquired language the way he did. I didn't fail with Shane (he acquired language!), but, as I said, it is easier to accomplish this task with a human embryo. With an ape embryo, this outcome is far from guaranteed, though it is very plausible and very possible. For both Teco and Shane, everything mattered: the culture and conditions surrounding development, from the time the embryo was formed as a blastocyst and began dividing and growing. You can speak of cultural milieu, or of amniotic fluid: it starts then, and it all matters. If you want to take it one step further, as many researchers do, then you have to examine the role of each particular gene or each amino acid on each particular gene. You need to consider questions like: "Did the experience that Teco's mother Elikya went through put this methyl marker on these genes and changed their patterns?" We have information from Derek Wildman indicating that, during Elikya's pregnancy, the genetic expression that was going on in the placenta was different from what is normally found in apes, and that it was more "humanlike." Although the parents' genes determined their son Teco, Elikya's placental development was also affected by other parameters—and, quite possibly, her rearing, as well as her current and previous environments. All of those mechanisms are operating. In addition to this are deliberate operations of the caretakers: what I chose to do when I saw Teco's inability to cling with his feet, or Shane's problem with vision because his tear ducts weren't working. I deliberately acted to facilitate cognitive and behavioral development, by design, in those two organisms. I interrelated all those factors to the language world in which they lived.

It is clear that we can have no definite answer to the debate about the respective roles of human nature and culture by pulling them apart. There is every answer to be gained by studying differential rearing conditions of various organisms. If we go beyond the here and now of those embryos and speculate, in a phylogenetic sense, about that which caused those embryos to appear on earth with the potential that they had as blastocysts, well, we immediately look for a Darwinian mechanistic explanation. I am not opposed to that. I just think that the Darwinian theory is far too simple to explain evolution on the planet. I do agree there is evolution: there is change across time. My own personal preference is to say that Darwin focused on the body as the thing that is being reproduced. I would focus on *behavior,* and not the body, as the thing that is being reproduced. Beyond behavior, what is evolving across time is the flavor and color of consciousness. Consciousness itself is what is evolving on the planet, and various forms of consciousness are evolving on the planet through behaviors and bodies—the growth of consciousness apparently requiring embodiment.

When I contemplate all I would have to do in order to enable a Teco to become a Shane, I already know that, no matter what, there will be many activities I will not be allowed to replicate with Teco: I won't be able to take him on a car on the road, or to let him play baseball. If he had every potential that Shane has, there would certainly be environmental limitations to what I could do, because of human laws. As for the limitations existing in Teco, I may not fully understand them. But, in parallel, I also know this: if I were to take a dog as an embryo, to do everything I possibly could to make up for the fact that it was a dog, and to treat it like a primate baby, I couldn't accomplish any of the major things I could create with Teco, who, very early on, put in a symbolic layer. As soon as I get this symbolic layer in an organism, I can accomplish a lot, for I am operating totally from there. With a dog, I would certainly get cooperation. A dog is not a wild animal I have to chase down all the time. It comes under some level of linguistic control. It can follow a point. This is why people are so interested in dogs: dogs can understand one thousand words; they can come into behavioral alignment with you. Many women carry puppies around, as though they were primate babies: they carry them in their sweater, or in their purse. They feed them with a little bottle. Those little dogs are raised like children, with often more money and time being spent on them than what most human children would get. Such owners want their dogs to be their children, and, by the way, there is little inhibition about it. Almost nobody says: "it is unethical to raise your dog as though it might be your child!" Dogs are being raised like baby primates. Still, no dog is recognizing itself in the mirror. Not even that! No dog is taking all the steps that Teco, Kanzi, or Nyota took, even in the restricted environment they were confined to. Dogs have access to the human and cultural world; they may even be "self-domesticated." Paradoxically, most dog owners would admit that their pet cannot do what their child can do, though they still assume that the dog's consciousness is like the child's consciousness. They have no problem with equating the two.

LAURENT. In *Animal Liberation,* Peter Singer was trying to avoid or minimize such differences. By going back to Jeremy Bentham's question, "Can they suffer?" Singer clearly states an equality of animals. He does not deny differences among species in terms of "self-awareness," "capacity for meaningful relations," for "abstract thought," for "planning the future," for "complex acts of communication"[7]—all abilities that are being debated in Steve Wise's ongoing legal actions in favor of granting habeas corpus to chimpanzees. But the equality in pain practically obfuscates the inequality of noetic capacities. Despite Singer's rightful attacks against Harlow and other psychologists creating psychopathologies in monkeys,[8] he is also—and unexpectedly—influenced by behaviorism when he proclaims a commonality in suffering. Years of behaviorism had reduced the psychological construction of humans to issues of conditioning, to action and reaction. In this regard, *pain* is construed as a reaction to a negative stimulus. I believe this emphasis on pain to be in strong resonance with another, apparently unrelated, discourse on "human vulnerability," a discourse tied both to scholarly endeavors and to the social talk denouncing victimization, "bullying," and any behavior that would "hurt" people. This whole phraseology is expanding and has clearly gained traction in the United States. I do not intend to deny the existence of oppression, of political "evil," or of interindividual harms, and I am certainly not defending factory farming or claiming that there is no call of distress in animals. Yet, I cannot help noticing that the rhetoric of the animal-rights movement ends up defining all animal life as *that which can suffer.* There, humans become the sentient beings (that is the animals) that are also sensitive. Their sensitivity is a psychological multiplication of harm (for they can abstractly consider pain, communicate it, anticipate it, etc.). At this point, my "consciousness" is either reduced to the expression of the harm I could receive, deliver, and potentially soothe—or to a specific "add-on" that cannot conceal that I am fundamentally animal in my suffering. In either scenario, the "spark" you wish to expound is simply out of the picture.

Within the animal-rights movement, we still find the imprint of human exceptionalism. This critique is not new, I know, but I am aware of no convincing counterargument. In the current configuration, the human being is still the agent that is supposedly going to edict and promulgate a law that will protect animals. Maybe against their will. Despite the proclamation of animal equality, one species retains the unique aptitude to organize the entire animal world.[9] In this manner of positing the debate, your own line of work can only make sense if it is travestied as a *preliminary* step toward the recognition of rights (of no suffering) of other animals. Now, what you said explicitly about our "lavender spark" is, in my view, incompatible with any of this. It is incompatible with the main goals of the animal-liberation movement, with the reduction of animalness to suffering, or with the tacit equation of all the different flavors of consciousness.

SUE. This is exactly right. In the paper I wrote for the upcoming legal conference I am speaking at in Minnesota, this is what I said.[10]

LAURENT. Okay.

SUE. I meant to send you the paper yesterday night but because of that Labrador tea you made me drink, I went right to bed before e-mailing you my text.

LAURENT. This Labrador tea was very efficient: it directly operated on your consciousness . . .

SUE. Yes! Maybe one way of summarizing what we have explained so far would be through the following flat-out statement: symbolic consciousness is about the self. Every body that becomes conscious is conscious of itself and may have the potential to become conscious of the self. To develop and expand in human culture, the goal of consciousness—that is first aligned with the individual self—is really to become more selfless and more aware that whatever one does is affecting everybody else, and to become more aware that consciousness should be about the greater good. Being aware of the self—or of one's own pain—is often thought to be selfish or animallike. Except that the animal is not aware of itself in the mirror . . . So, whatever is "selfish" or animallike in the animal, the animal does not have the self that we have, construct, and express with a languaged mind. The dog, once again, and as far as we know, doesn't have this self-reflexive consciousness, it doesn't identify itself as such, so to take something like pain that is an embodied experience and throw away everything else is to look for the lowest commonality between humans and animals. Every baby when it has diaper rash is going to cry because it has pain. If you poke it with a pin, it is going to cry because it is going to feel pain. I suppose even if you poke an amoeba with a pin it is going to retract.

LAURENT. Even the plant . . .

SUE. Even the plant . . .

LAURENT. I know there is a lot of discussion on the issue of pain, but one is usually led to posit a kind of arbitrary limit for the sign of suffering. For example, the plant does not cry. But this is not even true, because, as has been established, a plant can make noise, when it "suffers stress."[11]

SUE. Yes.

LAURENT. To go back to your last statement on symbolic consciousness and make a reference to my seminar that you attended yesterday, it seems to me that, throughout bifurcation and dialogic multiplication, "the self" could lead up to the possible creation of other selves, like those we find in literature—all the masks we speak through. The moment we separate ourselves from our self, many things are possible. It is like a proliferation or an expansion that is not additive or purely cumulative; it is like opening up another space or plane on which one can also live. Beyond the multifaceted individual self, beyond the group self, we find something I would call a disposable self that we create through fiction, through speculation, through the arts, and that we may decide to inhabit to such a point that it impacts our self. We partake in this other self than ours, though it is not meant to be ourselves, unless we are becoming insane, unless we are being possessed.

SUE. Is this disposable self a temporary self? Like a paper plate is a temporary thing?

LAURENT. It is a temporary self, but you can use it several times and it can be used by several different people with several different purposes.

SUE. You mean: my disposable self could be like your disposable self?

LAURENT. I mean: a disposable self could be found in a poem or a novel, and I am using it, then you can use the same disposable self, but it will not be *the same,* because you have your self and your group self that differentially relate to it.

SUE. Okay.

LAURENT. Putting on a mask on the face, like the bonobos do . . .

SUE. Is making use of the disposable self, in a way.

LAURENT. Yes. And we can go from pretense with a mask to embodying and reciting a lyric poem where I . . .

SUE. Become . . .

LAURENT. And feel and think the text.

SUE. This is very important.

LAURENT. All in all, from simple mirror recognition to poetry, we have indeed a wide array of potential expansions of symbolic consciousness.

SUE. And the essence of us is the ability to do this.

LAURENT. We might even say that.

SUE. But from whence this comes, we do not know. We know some conditions and experiences that elaborate this ability and others that diminish or even delete it. We can dispense with it, even in humans.

LAURENT. Yes.

SUE. This is what dehumanizing is all about; it is all about making that go away. But we don't know from whence the ability comes. I think we are about as close to understanding that as the ancient Greeks were. Unless modern physics comes up with something.

LAURENT. Well, I wouldn't place too much hope in modern physics in this regard.

SUE. I mean, we are talking here about energy. What is modern physics talking about? Energy. As you break down the atom into quarks and other things, you are breaking it down into particles and energy that, according to mathematical equations, are interacting with each other in certain ways to create and stabilize matter.

LAURENT. But the science of physics is still a human description.

SUE. Physics is more than a descriptive science, it is a predictive science. It also creates things, it doesn't just predict.

LAURENT. Yes, yes, but by saying that—

SUE. Now that we're at physics and that you have just picked up your rock again . . .

LAURENT. Yes, because I realized that I had forgotten about it . . .

SUE. Have you had any conscious intuitions? Is a rock conscious? Or has your consciousness picked up anything about the physicality or the energy inherent in the matter in the rock?

LAURENT. Oh boy, I'm afraid my consciousness didn't pick much.

SUE. It didn't?!

LAURENT. My semiautomated scholarly memory reminds me of this letter in which Spinoza is hypothetically granting consciousness to a rock that has been thrown into the air. "Don't you think that this enlightened rock would consider it has free will and is able to fly? And isn't it what human freedom is about?" Spinoza asks—if my memory is correct.[12]

[Sounds of protests suddenly come from the street.]

SUE. This is a demonstration of some kind.

LAURENT. Yes, this is a demonstration, but we don't know what they are demonstrating.

SUE. Can you tell what they're saying? There are a lot of them.

LAURENT. Hmm, it is already over. This was our interlude. Or maybe a prelude to a conclusion.

SUE. I'd like to say one more thing. When one has discussed and talked about consciousness until there is almost nothing left to say, and one is tired . . .

LAURENT. When one is tired, yes . . .

SUE. One can still sit and very effectively, without actually becoming more tired, experience consciousness in a very real way. You just have to be there and you can experience it in ways that are very difficult to put into words. Instead of having the dialogue we had, we could have said, "Let us sit in Laurent's living room for five minutes and experience consciousness; then, at the end of those five minutes, we will discuss what we felt." Would there have been any similarity in our self-reports? Possibly none. We could both experience consciousness in all its full dimensions and realize five minutes later that we had no overlap. So, when you think about it, well, it is hard to really know what consciousness is! . . . But, for each of us, it is very easy to experience it. If instead of just sitting here and experiencing consciousness, we were to do square dance together for the next five minutes and we jointly engaged in conscious embodiment, our verbal reports about our experience would overlap a lot.

LAURENT. But then, the overlap would be due to what we discussed earlier, and to some level of automation. I know there are anthropologists who currently try to use measurement methods to determine if, in attendants of collective rituals, the vast overlap in terms of verbal reports about their common experience correlates with what happens in their bodies (heart rate, tension, breathing activity, or even coordination of brain waves, etc.). It seems that, with very violent, spectacular, and significant rituals in particular, the synchrony is quite impressive—and we are not only dealing with "a manner of speaking" that would be common to members of a social group.[13]

SUE. You go off into a different state of consciousness. Many indigenous people will say that they go into states of feeling as one. When you live in a tribe from birth to death, you have such a commonality of experience. This, in itself, may enable you to synchronize with in-group members in a way that modern life, where we live separately in big cities, will not easily grant. The commonality of experience becomes the radio, the television, or Facebook, which is not the same as living together.

LAURENT. You're right. Now, in our societies, there are still moments where we are part of a large group and we feel there is some level of distribution of consciousness. In a massive rock concert, for instance, when you are really near the musicians.

SUE. That sound and the repetitive beat is traditionally used in indigenous societies to particularly induce that state.

LAURENT. Exactly. There is also this ritual component, either very directly (attending mass concerts becoming a rite of passage for teenagers) or a bit more indirectly (in the sense that, if you're a *fan,* then you have rehearsed, you know the lyrics and the music by heart). People cry. People scream. People move. This level of distribution might be very close to a collective trance, where you are suddenly being lifted up. The same phenomenon was exploited by the large Nazi ceremonies of the 1930s, where each dimension was ritualistic, and with the absolute necessity to raise the arm at some point so that all attendants would just be one. The modern army routinely plays with the same register: the warrior is being reconstructed as part of a platoon. A mob works similarly. In such instances, we do not automatically deal with the *abolition* of the self. Rather, the individual self is uplifted and appended to another *kind* of self.

SUE. You are being possessed.

LAURENT. Yes.

SUE. I am beginning to wonder whether there are actually two layers the way I talked about them.

LAURENT. I don't think there are only two.

SUE. There must be at least three. When the human self forms, there are at least three layers of consciousness. But I do think there are layers, and that when you are hypnotized or possessed, you are giving up one of those layers, or the conscious processing of one of those layers.

LAURENT. Maybe you are giving up a layer, or you are externalizing it. Or maybe—to change the metaphor—you are muting one voice, that a new voice could replace.

SUE. When people feel that another being is coming into their body, is consciousness existing beyond embodiment? Is it out there somewhere and able to come into another body so that it can express itself on this plane? If I ask Matata that question, I believe she would say yes.

LAURENT. I am sure that many people would say yes. I would have a harder time saying yes—or no . . . This might lead us to back to what I was hinting at about physics. Yes, physics is a predictive and, to a certain extent, "creative" science. It remains anchored in a human description. I am not sure that the progress of physics, in itself, could elucidate the sort of consciousness you are aiming at. When Roger Penrose is saying that consciousness, or the mind, does not function in a purely algorithmic manner, and that there is something beyond those operations, I agree. I may well have a different approach, a different goal, but I can sincerely agree with this critique. But when

Penrose, with the help of Stuart Hameroff and others, is trying to identify in the microtubule the material channel allowing him to reconstruct a quantum physics that would finally explain why consciousness and *noēsis* are not algorithmic—I believe he is mistaken. I am not hostile to the scientific enterprise of materialism, but, there, I can't help feeling that Penrose and Hameroff are fabricating the importance of a particular biological structure in order to make a point that was much better made, I believe, through the recursive consideration of Gödel's research on incompleteness.[14]

Then, let me make another observation. When quantum physics deals with "both this and both that" structures, this is certainly troubling from a scientific viewpoint. At the same time, that a thing be both itself and another, or that a state be coextensive with another—incompatible—state is everything but new. The novelty is that we'd end up with having a cat being both dead and alive by following a *scientific* theory. For the rest, non-explosive contradictory judgments are everywhere in our "symbolic consciousness." They happen in language all the time, and literature is particularly fond of them, forcing both semantic stability and logical consistency to mean, beyond the transparency of referential communication. Religious conceptions about witchcraft, ecstasy, and possession also dwell on the same construct. The macroscopic and physical world that we observe with our senses does not tolerate two different objects occupying the same space at the same time. But we integrate this paraconsistent property in our speech, in our social structures, in our self-reflection about feelings, in the very fabric of what you call our flavor of consciousness.

SUE. I see. You know, I have always thought of language as the playground of modern physics.

LAURENT. In the description of the science of quantum physics, I see elements that are very much in tune with what I find to be operating in language and in "symbolic" *noēsis*. We find a convergence between quantum physics and human language, in terms of "entanglement," of nonconsistent (though nontrivial) configurations, of nonalgorithmic operations. Let us set aside the idea that we would just have random or loose analogy between the two—or that positive physical determinations could be equally made on an electron and a poem. There are only two other options I can see. Either this convergence is casting some new light on a structure—or a law—that would transcend matter, energy, and our very own ideas, in which case physics would become the science of the unified reality of the real, that is, some metaphysics—or the convergence betrays something about our own understanding and our own condition as intellectual and conscious beings: and it mirrors our multilayered and discontinuous consciousness.

SUE. That may well be the case.

Postscript

As we were doing the final revisions for this part of the manuscript during the spring of 2017, both Sue and I read a very interesting article by Liangtang Chang et al. showing, for the first time, that with appropriate training Rhesus monkeys are able to recognize themselves in a reflected image.[15] This is an impressive result whose consequences and existence need to be examined. In an electronic message to me dated February 20, 2017, Sue wrote: "Monkeys do not engage in much eye contact. If you look at them for a long time without averting your eyes, they threaten you. Perhaps this is why they look at mirrors long enough to figure out how mirrors work but not long enough to learn that the face in the mirror is theirs—unless forced to do so by head restraint, as in this experiment [that is, the research conducted by Liangtang Chang et al.]. It might be that if several monkeys who had been taught to recognize themselves in mirrors raised offspring with mirrors around them, their offspring would recognize themselves in the mirrors without the training. It could also be that a few generations of this rearing—along with tasks that required following invisible displacements—would produce a group of monkeys that began to demonstrate causal exploration and tool manufacturing. Except for the Japanese sweet-potato washing example, there are no systematic cross-generational studies of the effect of a new skill on ensuing generations and on the group. While potato washing is an 'activity,' facial recognition in mirrors is not only an activity: it allows the monkey to reflect on the appearance of the self."

Sue and I briefly saw each other in Ithaca in May, and we exchanged on her work, this recent experiment, and the new state of university research in China. In a later message she sent me, Sue elaborated further: "We see that learning at the level of eye contact and gaze following is clearly *not* innate and therefore is not species-specific—and that eye contact, eye following and mutual attention orienting to specific aspects of the surrounding context are learned and become cultural instantiations which lead to: (a) co-joined perceptions, (b) common time-linked processing of those perceptions, (c) the formation of common co-conditioned patterns of response and internally based logic structures. This insight (that what we have viewed as instinct is learned through early eye contact patterns and co-joined bodily experience) provides the basis for beginning to scientifically demonstrate that behaviors which have been assumed to be based on 'instinct' and thus species-specific are instead *learned*—and learned through processes that include all of what has been called classical conditioning, operant conditioning, discrimination learning and concept formation in animals. It also provides the missing linkage between these forms of explicated learning and what has been called cognition by some and 'emergents' by Duane Rumbaugh. Once this link is in place, it lays the pathway for language to emerge if and when a second-order self concept arises or at a moment in time where the organism pauses in acting *as* the self and starts to engage in behavior *about* the self. At first, such moments are brief and fleeting but they begin to build, if (and/or as) these moments become endowed with symbolic reflection. This raises Deacon's question about 'the symbolic species': is the capacity to symbolize unique to humans alone? If not, what bootstraps it into a language? Such is the value of the animal language work. It is opening a door into the things which give rise to language—one of them being reared in a 'languaged world.' This kind of rearing—coupled with a large brain—allows for the opportunity to build a symbolic world that is initially co-constructed by behavior and dialogue, that is later internalized. This internal interpersonal dialogue later constructs its own *invented* internal world, and, through language, conveys that world to others who then may begin a sort of co-action which is based on another internal invented dialogue, becoming the seed of a new co-construction and culture. Hence, a reality arises from mind itself and is then co-formed through the vehicle of language. The mirror recognition study is thus a key link to bind human and animal learning into a common progression. It is interesting that this new study with monkeys comes from China where, from what you are telling me, there is an intent to bind humanities and technology together and where many aspects of a culture are opening up to a new language and putting aside some of the former ways while keeping others."

On Language and Apes

In the Great Ape Trust of 2010 or 2011, functioning and easily accessible computers were scarce. Still, and by design, written language was everywhere: symbols were posted on each room, the T-shirts of the caretakers had all the lexigrams printed on them, and paper copies of the panels were available in all enclosures and in the lobby area. The apes could constantly use their language in their informal exchanges with the people working around them, without the need for a computer. For actual, reciprocal communication to occur, humans had to be proficient enough with hundreds of signs, and to remember where the words appeared on the panels. In the absence of a synthetic voice, a good knowledge of the spatial distribution of the signs was a sine qua non for any person to grasp what was said. In my view, the problem was never so much the level of oral comprehension of English in the bonobos (it was very high), but the ability for nonnative speakers of Yerkish—like me—to at least readily recognize the words selected by the apes. At the end of my first visit, a computer scientist then working at the Trust gave me a software for Yerkish acquisition. I trained myself, and, in 2011, I was much more at ease. With a copy of the lexigrams in front of me, I could rapidly identify the words that were pointed at. This is when I became "ROB" for Panbanisha. When referring to me, she stopped using the generic term "VISITORS" and simply selected a name that was present on a panel but was no longer associated with someone from her environment. Yes, I am saying that Panbanisha was able to distinguish between a common and a proper noun, and that she thought of individualizing me through the "ROB" key.

This same morning, an exchange occurred between Nyota

and me. We were on the same side of the lab, just separated by the mesh, with no other human or ape in sight. He was playing with an old garment, and I was observing him. He took his paper keyboard and glanced at me directly, as he usually does to begin an interaction. Then, he pointed at "SAME," looking at me and clearly expecting an answer. I was quite surprised, for, so far, I had not seen Nyota employing an abstract term like "SAME." In the past, he mainly asked me to bring him food ("BLUEBERRY" being a favorite), or to play with him (in selecting "CHASE," "TICKLE," and "MASK"). I could now say there was no ambiguity about the lexigram: it was definitely "SAME." So what? Nyota was a bit surprised by my surprise: he continued to look at me, still playing with his torn piece of clothing. I had to ask him orally what he meant ("I'm sorry, Nyota, I didn't understand, what are you speaking about?"). To which he responded with the symbol "SHIRT." I finally noticed that the garment in Nyota's hands was a yellow shirt with a square pattern—in this similar to the one I was wearing that morning. "Oh, yes, you're right! We have the same shirt!" I said. Approvingly, he nodded.

Having these memories, and many others, I entered a new conversation about language with Sue the day after we exchanged on the flavors of consciousness. We gathered a few arguments that we considered briefly over lunch at the old hippie vegetarian restaurant. Then, as often, in a sort of intellectual preparation, Sue asked me to summarize twenty-five centuries of philosophical thinking. My professorial self obliged. Thirty minutes later, Sue said: *"I hope that's left in the book."* (It is not.) From there, we moved on to the issue of syntax, and the inflated importance grammar has acquired in the last decades, through the expansion of generative linguistics. Wasn't it the case that Noam Chomsky, as he himself stated, had been influenced by Descartes?

LAURENT. In the few texts where Descartes speaks about language, he tends to take a stance that is not exactly the one we would expect from an author who is so carefully innovative in his own style, especially in French. In some respects, despite his opposition to Scholasticism, Descartes is repeating Aristotle's ambiguous position, which was both encouraging strangeness in writing and claiming that semantic ambiguities were just a halo. At any rate, the Cartesian method is also able to make verbal discriminations in such a way that to each concept one word (or one set of words) will be related, making a discursive demonstration intelligible. This quest is very widespread in seventeenth-century France. A bit later in the century, one even finds *poets,* like Nicolas Boileau, who claim that words do not need to be equivocal and should proceed from mental clarity.[1] One could almost argue that the general emphasis on structures, rules, and clarity accompanies the gradual rise of absolutism in France, culminating with Louis XIV. This would be an analogue to Erwin Panofsky's reasoning on *Gothic Architecture and Scholasticism.* Just as symmetry rules over French classical architecture or the art of gardening, words are supposed to be well aligned with concepts. Sense stems from clear and regulated relations, and all this expresses ideas. For the ancient Greek thinkers who were advocating a positive use of language, the goal was rather reached by some constant definitional refinement in accordance with logical rules. Wherever they stand on this issue, most—if not all—theoreticians would feel themselves to be bound by both of those aspects, at least to a certain degree. Even I, who consider equivocation and trans-consistence to be crucial, would not state that reasoning should proceed without any care for refinement or for epistemic regulation. Now, while Descartes is skeptical of the actual implementation of any new artificial language that could be common to all mankind, he also considers the potential benefits of a universal idiom. Since each language has two essential sides, Descartes says—namely, *word meaning* and *grammar*[2]—one should first dispel the confusion of word semantics and arrive at the "clear and simple" ideas that vague lexica would normally represent. This is the job of "true Philosophy."[3] From there, one could create a methodical "order" linking concepts through unambiguous words.

SUE. So, Descartes is really making a statement about how language should operate.

LAURENT. Yes. Cartesian analytic geometry identifies coordinates along axes and situates points and objects according to numbers. The project for a new "universal language" is envisioned through the same method. Descartes explicitly compares the new "order" of word concepts with the faculty of counting.[4] Once one clear concept corresponds to one clear term, what you need to do is to link those ideas together. The phrase "universal grammar" appears in the same letter by Descartes that I just commented upon.[5] It needs to be said, however, that, in the letter itself, Descartes clearly says that grammar could only be *made* universal through artificial means. He is not suggesting *at all* that "UG" would ever be found in the "natural languages."

SUE. In trying to find the deep grammar, Chomsky is doing a kind of analytic geometry of language.

LAURENT. Yes! Chomsky could even claim that semantics is an interesting topic, but that grammar should be dealt with in the first place. In contrast, some analytic fixity of meaning had to be *presupposed* in Descartes's geometrical emphasis on alignment—and in any similar endeavor. The emphasis on grammar as what *allows* a clear usage of words is something different.

SUE. Then, understanding the famous sentence "colorless green ideas sleep furiously on the grass" is similar to computing the hypotenuse of a triangle when you have the other two points. You can figure it out. This is what grammar is all about.

LAURENT. Yes.

SUE. When people finish each other's sentences, or when they are talking in a very colloquial style—*and they . . . what about . . . oh, man!*—if you just look at what speakers say without picking up the whole pattern or understanding the context, you, as a Chomskyan linguist, could not make much out of it. Thus, linguists do not pay much attention to such phenomena. Because this wouldn't fit their idealized view of what language should be. Unless there is something beyond what is mapped out, something so deep that it could handle this ordinary, everyday language that doesn't really fit. A new word can pop in, and its meaning can be understood in context, even though it cannot be defined in the dictionary and is not associated with any particular concept. People who are not really using grammar that fits any concept of grammar somehow still make meaning. So, Chomsky is looking for . . . let us call it: the operating system of language. And he is influenced by Descartes's analytic geometry, because there is a geometry in syntax, and even if you express it through a mathematical equation, it is still not the equation that has the power: the power is the geometry, which is a n-dimensional thing that is projecting onto the world. Chomsky is looking for that analogue in language.

LAURENT. Yes. In this regard, the emphasis on the symmetry or antisymmetry of syntax is telling.

SUE. The only way new meaning is arising is if you have these concepts and words out there that form some kind of meaningful whole, like a pyramid forms a meaningful whole when you put these lines together. It is far more than the trajectory of the given lines. Is this what Chomsky is trying to say about language, that, by studying the structure of grammar, he is really looking for the system that takes bits of meaning people think of as symbols and that orders them? Does he seek the alignment that is formed by the structure of the language rather than the alignment between the concept and the word?

LAURENT. I would say that you develop this interest in the second (grammatical) alignment, because you have presupposed the existence of the first one (between words and concepts). Admitting the presence of this first alignment is already a huge presupposition, which was taken to a grotesque level by Jerry Fodor and his chimerical "language of thought."

SUE. Let us talk about this first alignment, then.

LAURENT. Agreed.

SUE. In the early period of the "ape language research," the Hayes, the Gardners, and Herbert Terrace seemed to consider that the work of teaching a chimpanzee was just to make an association between an utterance—or a sign, or a symbol—and a concept. Like in paired associate learning. You show to an agent a wide variety of novel shapes, you attach a different sound to each of those novel shapes, so you have given "this word" for "that concept" (that is a shape). And you have a one-to-one alignment of "this word" and "that concept." When people began to try teaching words to chimpanzees, they first made the assumption that the chimpanzee had the concept of, say, apple or banana. Then, they assumed that the "word" *apple* or *banana* (it could be a shape, a gesture, a sound), as it was paired with one particular banana or apple, could—by some sleight of hand, by some miraculous process called associative learning—come to *represent* any apple or banana, and could stand in lieu of it in a communicative event. The idea was that, by paired associate learning, you had the ability to reference, *for free,* and that, *for free,* you also got the ability and the knowledge of using reference in an intentional communicative event. You didn't need to *teach* any of those two abilities. All you needed to teach was an association, then you would look if those associations fit a grammatical structure. This is where Terrace comes in. In fact, Terrace isn't even dealing with the idea of whether you actually have true reference, and whether you actually have referential, intentional communication. He immediately leaps to syntax, which is the structure of how words like *apple* and *banana* were put together, and he concludes that Nim isn't putting these words together in a proper kind of sentential structure. Therefore, Nim cannot "have," within himself, "language." But no individual can "have," within oneself, "language," if the latter is referring to nothing. If language is reduced to paired association, you can *produce* language but you don't "have" it, you don't "own," nor "possess," it. In such a situation, if you don't have reference to begin with, there is no sense in trying to figure out some inner structure establishing how these paired associations are put together.

Let us say we do the following experiment. I give you a list of paired associates with a whole list of nonsense objects and a whole list of nonsense syllables. I ask you to memorize them, and you do it, maybe even in one or two trials. Every time I show you this strange drawing, you go: "Blelelel!"

LAURENT. Blelelel!

SUE. That's right, Laurent! You are very good at this task. I give you a blind test, I show that I am not cueing you, and you reach a level of 100 percent accuracy. Now, I wish to look at how you put all this together. First of all, I must create a situation in which you have to put those syllables together. But, so far, you have only learned a lot of associations. Unless they are inherently referential and inherently communicative, you have not learned more than a series of associations. Both Terrace and the Gardners skipped over that phase. Moreover, I guess, they had an implicit belief—they never stated it—that chimpanzees had the ability for symbolic reference, and for symbolic communication. They just assumed it existed, within the chimpanzee, although chimpanzees were thought to be dumb and to have *none* of the true essential components of language: none of them had been identified by anyone in the wild. So how could one presume that those elements were already there, *in* the chimpanzee? Even the simplest things like *pointing*—that many development psychologists consider to indicate the ability to make a statement in a child—Terrace and the Gardners don't address them. There arose this huge disappointment in their work, and questions like "Can a chimpanzee imitate?" (which has occupied reams of paper), or "Can a chimpanzee point?" (which still isn't agreed upon today). But there was a huge problem in how the question of language was addressed. Terrace deflated the whole field by arguing that Nim did not have grammar, because he repeated things. As for Washoe, the Gardners, when they reported her utterances, did not accurately record sign order, because they did not think it mattered; hence Washoe reputedly had no grammar. Except that, in many human languages, word order is not the key. You can use words in many orders, sometimes in almost any order. You may use a slightly modified version of that word, through morphology, but maybe there was a slightly different version of the signs the apes were making. It has not been studied. No one has even looked at the films the Gardners have taken to answer this.

The basic question about language becomes, Is this "animal" able to engage in reference at all? If it can, it knows that this symbol can do away with this actual banana, and this word itself can cause it to think of a banana. As it becomes referential, a symbol could be manipulated by my mind, and, in a certain way, it is freed from this or that object. I can intentionally do the same with others, who know that I am doing it, and I know that they are doing it: we know we are in a process of doing it, and this is why we are using these symbols. We are using more than one symbol, so we can modify the meaning of this word. If I have just one symbol, how many different meanings can I give to "apple, apple, apple, apple"? But if I can bind *apple* with *no* and utter "no apple," I have suddenly created a hugely different meaning. I have negated what I have just said. I am sure that all the thinkers you mentioned in the first part of this dialogue, even if they didn't discuss these issues, had grasped those aspects of language in some deep manner.

LAURENT. Yes.

SUE. But virtually none of this is mentioned or considered when the debate of "ape language" emerges. For the first ten or fifteen years of modern ape language research, throughout the scientific community, the assumption is that nothing is missing in the way language is approached theoretically and taught practically. All you really have to do is to look at what the ape says, and then, by looking at what it says, without attending any of the so-called peripheral aspects, you will be able to figure out if it has language. This is where, much later on, Pinker's critique of my experiment comes in.[6] The reason his statement comes in is that all of the real dimensions of language had been preliminarily cast aside. I would like to hear you speculate about why this happened. When I came into the field of "ape language," it was very clear to me from the beginning that those dimensions were missing and that we needed to understand them. It was overwhelmingly self-evident. I can't figure out how they went missing.

LAURENT. In many situations, when you have something that could be called an intellectual tradition, the theoretical framing that was once created in the moment a conceptual problem was formed will tend to be kept for centuries, either in an intact or degraded form. Absolutely brilliant minds could consider many interesting consequences and rearrange the blocks—but within the given frame. If I distance myself from the framing, I can still admire the brilliance and the virtuosity of the practitioners, but, already, a large part of my own interest has vanished, because it was so conditional to the frame itself. This is how many thinkers within medieval Scholasticism or analytic philosophy can display some impressive intelligence and knowledge of "the field," while being unbelievably tedious to read, and, in the end, terribly negligible. Because they just stay in the frame. *Framing*, while being theoretical at its core, also exists in nonscholarly endeavors: it intervenes in the structures of societies, religions, etc. Undoubtedly, at a given instant, you may have different framings that are being used, but there is always a dominant one.

The dominant conceptual framing of *language* in classical philosophy, then in the Middle Ages, is not about everyday conversations. It is not about the origin of speech. It is not about the transformative power of discourse. All those three aspects do appear in the Greek and post-Greek traditions: Herodotus narrates the story of Pharaoh Psammetichus, the founding father of experimental psychology, who isolated human children from birth in order to determine which language they would speak "naturally" (this was Phrygian).[7] The language of poetry and theater has been perceived as being a tool to shape minds and behaviors; and the discrepancies between "normal" speech and other uses of language have been noticed widely. But, in philosophy, the majority position about language is tied to technical problems about the verbal status of truth and the relationship of what is said to what is thought. One big concern, for instance, going from pre-sophistry to nominalism and beyond, is about the combination of *being* with *not,* and the discursive creation of *nonbeing.* Does the latter exist? Is it a mere effect of words? And *is* there even something like nonbeing? etc. We are a bit removed from "apple" and "no apple," despite some commonalities. In the Greek scene, "everyday language" is certainly envisioned, but it is construed as a step toward a higher speech order—be it poetic, philosophical, comical, political, sophistic, religious, or even mathematical.

Now, we skip a few millennia. Many theories have been developed; the quest for the origin of language has become a European obsession before being officially banned for decades as a serious scholarly topic, and new insights in biology have appeared. But maybe the Gardners and Terrace are still roughly operating within a very traditional frame that has been both complicated and adorned over centuries of debate. What they can see through the window is still completely constrained by the frame. They may, unknowingly, try to give words to a chimpanzee, while retaining one possible definition of language that is mainly the product of technical controversies about being and nonbeing, or on the reality of reality. Terrace and the Gardners could have found many other ideas about language in the minority traditions that were much more propitious to the questions they thought they were addressing. But, operating as standard scientists, they retained the dominant qualification of language that most scholars would examine if they wished to enunciate "truths."

Remember that you asked me to speculate! . . . I would not "bet the farm." Still, I would sum it up like this: out of laziness, conformism, lack of epistemological concerns, or a combination of the three, the researchers you mentioned adopted, for their own experiments with apes, a long-standing framing of language that was not the only one that was at their disposal, but that just happened to be the dominant one. Unfortunately, this framing never had anything to do with the question of even how *human* infants could get language. So, it proved wonderfully problematic with chimpanzees.

SUE. I understand that it had nothing to do with how human infants get language. I don't want to fault the Gardners so much for that. Let us say they are only psychologists, and they don't have this background in linguistics or in philosophy.

LAURENT. This would not prevent them from "being framed."

SUE. Yes, the world may be framing them. Yet, as psychologists, they know what a paired associate list is. They know that humans with full linguistic competency, as they learn a paired associate list, just learn those associates. The Gardners know that these behaviors can be "conditioned," without any idea of reference whatsoever. So, their main approach was: If we teach Washoe these paired associates, we will condition her to do this, then what we need to determine is if Washoe, when she is being shown a slightly different picture of an apple, will produce the associated sign that goes with it. Now, with regard to *apple,* you can have an apple with different orientations, you can have a fourth of it. I want to go back to the example of this person who has been blind most of her life and is being given the capacity to see by modern techniques. She can now see something like an apple: she already knows what it is and how to represent it linguistically. But if you present the apple in a slightly different orientation after she learned to "see" it and relate it to the word *apple* and to her concept of it, she does not know what it is. Because her visual system has not made the transcript. Word knowledge plays no part there. What the "mind" is doing is not identical to what the perceptual system is doing.

I don't know about Washoe because she wasn't tested on this, but I can tell you that, with Sherman and Austin, when they could use a symbol for an apple, the first time you presented them a cartoon for an apple, they didn't know what it was—Kanzi did. Sherman and Austin couldn't make this kind of a perceptual transmission. The more important transmission for the Gardners, and the one on which they focused all their writing, was a word like *meat.* If Washoe could identify all these different objects as *meat,* even though she did not initially learn to associate the sign with all different kinds of meats, if she could generalize, then Washoe was able to go beyond the original paired associate context she had been taught. Therefore, she had the capacity for reference. If the paired associate *enabled* reference to emerge in Washoe, then Washoe already had some reference, and she already had symbolic capacity. She already had all the traits that the human child needs to go on with language in order to become far more competent than she could be. If the

Gardners were correct, if Washoe could generalize across meats, she could construct the symbolic layer up there in the world. This could only mean to them that Washoe was operating beyond conditioning the way they had traditionally observed it in pigeons, rats, pigs, monkeys, or any other species psychologists have studied.

Since this time, other people like Ed Wasserman have given pigeons very complex pictures that the birds might experience in their daily life, such as the image they might see when they fly over cities. Scientists have given the pigeons multiple paired associate training with a wide variety of pictures, and they claim they found generalization in pigeons. Does it explain Washoe's behavior? No, for the ability to generalize and not have one word go with one concept, which comes out of the philosophical framework, is not equivalent to the ability to be symbolic. Which is the question I initially asked. How do we determine, within an acceptable theoretical framework that is saying there should be a one-to-one mapping of concept and word—a position many psychologists have accepted—that the mind can go beyond this, and that it always needs to go beyond one utterance and one concept? We are not paired associate creatures. And how could I try to teach these two beings, Sherman and Austin, who had no real social background, who were reared in a nursery, who came out to me rocking and poking their eyes? How could I teach *them* language? What frame do I start with? And whatever frame I pick, how do I know if I found symbolic reference? How can I communicate with the world about the abilities of Sherman and Austin, if the world retains the conceptual framing you spoke about? How may I go beyond my "subjective" claim that there *is* reference? As a psychologist, I am taught to do this through presenting data. Good. Now, what kind of data would be demanded of me? If I look at my predecessors, the Hayes, the Kelloggs, the Gardners, Terrace, and Premack, they don't give me one clue as to what is needed to address the issues of symbolic reference, of mutual symbolic engagement in a series of dialogues with multiple uses of symbols that are connected to each other through something that Chomsky is calling grammar. Everybody tells me that if I can just find grammar, then it is real language. As I began working with Sherman and Austin, this was so far beyond the pale for me that I simply abandoned considering what Chomsky was even talking about.

LAURENT. I would like to repeat that the conception of what we could call "purified language," where one given word or word group corresponds to one clear concept, is really introduced by most thinkers as the way things ought to be—but are *not*. When Socrates is meeting with regular people, with sophists, or with philosophers, all the participants in the dialogue first need to come to an agreement about what they wish to discuss. In the original framing of language in philosophy, language has to be purified, *because* it is *not* pure. Within the consistency of this tradition, the whole project is to repair the undeniable default. In the modern era—and this is possibly due to success of the formalized notations of mathematics—more and more authors will think that "natural languages" are either already pure (deep under the surface) or just impossible to purify. When Frege introduces propositional logic in what he calls *Begriffsschrift*—that is, the pure, direct, *writing of concepts*—he is justifying his invention by an irreparable deficiency of human language. Tarski will be close to this position.

Another thesis argues that because one can somehow purify language—which was the old promise of dominant philosophy—then language, ultimately, or in the last analysis, *is* already pure. Thus, we don't need to care so much about "surface realization," or "noise," or "inconsistencies," indeterminacy, and vagueness. What a strange inference! The move is exemplified by a very odd shift in Chomsky. In his minimalist period, he argues that his whole enterprise will *prove* that language is perfect. A few years later, you can hear him saying in a talk that he will begin with a "methodological principle" enunciating that "language is perfect."[8] What used to be a point of arrival he had tried to reach now works as an axiom. If one cannot show that language is perfect, it is certainly better to begin with the undisputable assumption that it is . . .

SUE. And if you are Noam Chomsky, you can do that.

LAURENT. Exactly. Admirers will say that Chomsky revised his research program: this is what a real scientist should do. In the meantime, what can we think of someone turning an impossible conclusion into an underlying assumption? This is one way of spinning things around.

SUE. And it is language that allows you to do this.

LAURENT. And because language is logical, then, in a sense, you have almost proven what you said by saying it.

[We both laugh.]

One of the paradoxes, of course, is that, both within philosophy and outside of it, one always found some raging reaction against the ideas that language would be pure, or purified, or even purifiable. We can say that the majority approach to language is the outcome of many simultaneous and successive simplifications, and that, when we arrive at the suggestion that only grammar matters, or that it is perfect, we are dealing with a caricature of caricature. As I said before, this does not prevent brilliant intellects from operating within the frame and spinning things around very artfully, like jugglers in a circus. But while the official topic for discussion is "language," its conceptualization is divorced from its use, from its acquisition, and from its transmission—and from what speaking allows to achieve and build. The concept may be *interesting* (within its own frame, in particular); under certain circumstances, it might even be locally true. But since language is far more than the object of generative linguistics, what the latter could say about an utterance in an ape is generally useless.

SUE. Then, there is another wide misconception that somehow, in the field of "ape language research," we are trying to deal with the *origins* of language. Some scholars—Sebeok is one, there are many more—have taken a position that is similar to the pronouncement of the Linguistic Society of Paris in the nineteenth century: It is bad to do this. We don't really know why. Is it bad in the same sense it was thought to be when the Linguistic Society considered the debate? I don't see this as being the case with ape language at all, for it is an experimental discipline. Aside from taking a lot of time and effort, there is no reason not to do it, unless you want to ethically say it is wrong to do it *with a chimpanzee.* If you are taking that path, then you are making a statement about this other species in saying that, had it language as we do, this would not be a good thing. But then why is it a good thing for us? Or you are stating that this other species can never have language. But how do you prove it, if you are not testing it? Stating that if humans were meant to fly they would have wings, or that if chimpanzees were meant to talk they would be talking, and since we can't see they are talking, they must not be talking—all this is another cop-out, and an insignificant way of attacking the work. It is a trivial attack. It may be highly effective politically, though it is anti-intellectual. I hardly consider it worth answering, because we have to step outside of the academic intellectual framework of the experiment to answer this attack.

So, we need to go back to the experimental framework. In this paired associate learning, how did we get Washoe to "produce" reference? Did she just learn a list? You can't say "Washoe, I want you to learn *x*." You can't give her a grade or money, the way a psychologist would do with a college student. You have to reward Washoe every time she does *x*, but *how* do you reward her? Do you pat her on the back, or do you take her out for a walk? Do you give her a piece of food? The thing that seemed to make the most sense to the students who worked with the Gardners—and the situation was identical in the labs of the Hayes and of Terrace—was that if you want to teach *apple* to a chimpanzee, you cause him or her to sign *apple,* then you give him or her a bit of apple as a reward. Then, what Washoe or Nim are learning is not really a paired association. Washoe is not exactly learning to match this geometric configuration with this utterance or this gesture in the presence of a particular object or picture thereof; she is learning that, when she gives this utterance or gesture, *she gets to eat an apple.* Instead of having a two-pronged paired associate, we have, in fact, a three-pronged association: this apple, the symbol, the causal relation between sign making and having food. So, what is this "symbol"? Does the sign referentially convey an internal desire ("I want an apple")? Is this gesture a good trick that gets you an apple, in the absence of any referential intent? This became, and this still remains, the crux of the problem in ape language.

The answer to the problem lies in something that is still very much embedded in this "one word one concept" approach, but with an important twist. Evaluating what is going on *in* Washoe's mind, body, brain, self (if her self exists), as she makes the sign for apple, cannot be directly assessed by giving, or not giving, her an apple. Within the experimental paradigm as it was laid out, it is difficult to make a correct assessment. Furthermore, if Washoe gets no apple for some time, does the ability of whatever gesture or sign she made to represent *apple* diminish or dissolve because it no longer results in having an apple? If it is truly referential, no, it never goes away, it can be used for the object, and in its absence. In contrast, if you start to make the sign for *apple* and you do not give Washoe an apple, then, within two or three trials, she will be signing *banana,* just to try if another sign would work and get her food. From there, Washoe might proceed toward confusion, before grasping one-to-one correspondences between signs and desires. Psychology would be reaching, in an experimental way, for the purity of language you mentioned earlier.

Now, we are in the real world and in messy life, trying to get along, to get Washoe fed and diapered. Washoe wants to bite you, she wants to tickle you, she wants to go outdoors, she doesn't want any more apple though you need to test her for *apple.* And not only *apple,* you have to test her for at least twenty other signs, because, in the psychological paradigm, you really need a larger *n* to analyze your data. At this point, you can't even begin to seek language purity with embodied communication. You are beginning to very operationally grapple with *what language does,* and not with what language ought to be. What is language doing, in this context of social interaction between me and this organism, that I could not possibly do if there were no language between us? As a parent, you become very quickly aware that if you can do some things through language with your child, it is much easier than if you were trying to do without. This period of realization that every parent goes through is *very* short. At first, the baby can't walk or crawl, so whether it has language is marginal: you are not having to do a lot of negotiations with an organism that does not have the ability to go out there and say something. You are really in charge. Then comes a transition period, when the child has some modicum of language. You don't know exactly how much language the child has, and what you can say to it that will, or will not, be understood or taken into account. As a parent, you are in a gray area. This gray area does not last long. By the time the child is generally two or three, although sometimes it may run up to five or six, it clearly has language. If I am asking the child orally not to do *x* or to do *y*, you know it gets the request, and I am starting to negotiate with it on a linguistic plane.

As soon as the child is functioning in the human domain on the symbolic plane, it doesn't matter so much if it does not have all the pieces. When a child does not have enough pieces to be functional, this is different. I am thinking of the kids raised in Romanian nurseries at the end of the Ceauşescu regime, or of Sherman and Austin, who had been reared in even worse conditions when I got them. And they were chimpanzees on top of this! . . . You go through this difficult period with those organisms, ignoring if they even have the capability to be symbolic or referential. And every day, you are trying. "Here is the symbol. Here is the banana. Here is the symbol. Here is the apple. Here is the symbol. Here is the cup. Here is the symbol. Now do you know which symbol goes with apple and which one with banana? Oh, you don't know! We'll have to give you more reinforcement and see if you can learn." You see the chimpanzees mainly rocking; you are not sure you are getting eye contact. The same would be true of an autistic or retarded child, who hasn't gone quickly to the language plane. What needs to be accomplished with this organism in order for it to reach the symbolic is not what a parent would do during the small, so-called critical, period of language acquisition in children.

For most people who are trying to work with apes and teach them language, this period of criticality corresponds to the transition from the sweet chimpanzee baby you can carry around to the young ape that hops and climbs. At first, you can put clothes on the baby chimpanzee, and have it with you, because it is so cute! Then, about five or six months later, the ape begins to toddler walk, already crawling on the floor. It is still cute, and you can keep up with it, but once it reaches the age of one year and a half, it is impossible to keep up with it. The chimpanzee is three-dimensional now. It is climbing on the fan and on the lamp, it is everywhere. It is up there on the shelf! You can't even get up there to get it down: it stays up there and it looks at you. It is not being bad, it just wants to know what you are going to do! As a human being, you don't come physically equipped to deal with this situation. The only way you are going to be able to deal it is at the symbolic level.

Terrace's reported way of dealing with it was to teach Nim the word *no*. Isn't it simple? *No* is associated with the negation of everything else in the world: *no apple, no run, no eat, no on the shelf.* So, you just make an association between *a behavior* and *no,* then Nim will have to stop. Terrace tried to teach Nim *no* by drawing a line on the floor: Nim is on one side, and Terrace has instructed his assistants to teach Nim the word *no* (Terrace himself didn't do it). How do you teach the word *no* with a line? By picking up Nim and putting him back every time he crosses the line? Is this a reward? At least, if you are teaching *apple,* you could give Washoe an apple. What if you are now trying to teach Nim *no*? What do you give him? Do you give him an apple because, when you say *no,* he doesn't cross the line? Then, does he think that he gets an apple if he does not cross the line? Well, yes, Nim did think that! In Terrace's movie recordings, such lessons lead to scenes like this: Nim is trying to do something, like eating an apple that he just got as a reward, and the teachers say "no apple" and push Nim's hand down like "no! no! no!" It suddenly becomes a very negative, really affectively undesirable state of affairs. The problem is no longer whether Nim has the cognitive ability to make these associations or not. The problem *for Nim* is the one many autistic and retarded kids encounter when they are doing things their parents don't want them to do. The mind can't acquire anything in that negative emotional state. Yet, this gets turned into a "cognitive" question by Terrace, who is seemingly unaware of Nim's emotional state and of the importance of the latter for acquisition. Of course, for the psychological paradigm of conditioning, emotions are not to be mentioned, they might be left out. A scientist is not supposed to say that, as it manifestly appears on the tapes, Nim isn't learning because he is unhappy. The mind, however, needs to be in the right kind of state for the cognitive system to pull in additional information. Everybody would grant this to a human child, even though many practitioners of modern ways of working with autistic and retarded children still believe in the need to constrain the difficult behaviors in the child by shaping it through some similar technique. Terrace hoped he would achieve this, because he felt himself to be the ultimate shaper. Instead of getting the symbolic layer in and of allowing the joint integration of behavior to exert control, one tries to "shape" the behavior. As if by picking the child up and putting it down, you were actually shaping how the child is sitting down! . . .

I suggest that the problem for the world of ape language is not so much the cognitive capacity of the ape but its physicality. The physicality of the ape is so different from the one of a human! Still, there seems to be no willingness to address these questions. For example, on the part of Mike Tomasello at the Max Planck Institute, there is no willingness to even potentially address these questions. In Japan, on the part of Matsuzawa and Idani, there is willingness to interface with the ape as well as some recognition that this interface might be necessary to co-build a symbolic layer. But they don't ever speak of this in their writings. They generate literature through conditions that may involve the interface, but, whenever they speak of it—as they have done in private conversations with me—this is viewed as the art of being a chimpanzee handler that you get by practice and don't discuss in scientific papers.

If we don't put language under an experimental protocol of some type, I don't see how we will ever understand language, if it is possible to do such a thing. If we conclude that it is not possible to understand language from being within language, I suspect we are stuck in all of the philosophical conundrums you have described so eloquently throughout this dialogue up until now. I see this as the only way out of the study of the appearance—not origin—of language in another species.

LAURENT. New questions, or new reconfigurations of old questions, were certainly created by the ape research, but, then, the experimental setting can almost self-erase, if one neglects or conceals its "preparation." It sounds very weird to me to consider that the preparation, which includes the emotional state, would not be officially part of the experiment.

SUE. It is a part of it—but it is the part that, as a psychologist, you are not allowed to discuss. The only way you could *legitimately* discuss it from within the discipline would be by controlling the variables and having somebody taking data and watching you as you prepare to do the experiment. In other terms, the scientist would need to become the object of the experiment of others. This would methodologically make it possible and legitimate, but this has never been done. Every time I tried to go there, I faced tremendous resistance.

LAURENT. Scientists presenting experimental work are expected to include the description of their preparation. In physics, experiments could be dismissed, not on the basis of their data but because of the way data have been generated. However, I saw in reactions to your work that the part of the preparation dealing with the emotional state or the physicality of the beings participating in the experiment is often considered to be either subsidiary or insignificant. This might be a fault of experimental psychology, where practitioners routinely, and wrongly, dismiss the importance of the background of their subjects (the American prisoner, the male undergraduate, etc.). Yet, in the kind of research you have done, beings have been "produced" in a lab (which is not the case of the humans that are tested).

SUE. Yes, and this preparation is the entire life—it is the whole thing. However, as a discipline, psychology never looks at the whole of life. Long-term life-based studies are exceedingly infrequent in psychology. Cross-generational studies are almost absent. When they exist, they look at only a little part of the life which often could be measured by a questionnaire, or maybe now with a little bit of videotaping. Studying what even one baby does for one whole day, you don't do it. This is why some of the greatest work has been done by people like Piaget who were actually around their own children for all of their lives. By seeing that child every single day, they were looking at its cumulative development. But, typically, if you are a psychologist, you view this world as existing out there in the same way a sociologist might view the society as existing out there, or in the same way a chemist views the chemical world as being independent of his studying it. You go out there, you take a slice of the world, you come back, and you analyze that slice in your lab. This is what the people who have now taken control over the life of Kanzi and his family call "real science," when they dismiss out of hand the entire importance of the rearing paradigm itself, while still claiming at the same time that they follow my "research trajectory."[9] I am sorry: if you study what a bonobo is learning or acquiring, you have to talk about all the preparation that went into *this.* As for the larger preparation with the bonobos, it has no clear beginning, no clear end. Every bit of your preparation is affecting the outcome. You are never not preparing. I was most acutely aware of it with Teco. There was not one second with Teco that was without the knowledge of where this behavior at this moment of time was intended to lead to many years later. Not one second. No downtime.

LAURENT. I very clearly understood it the first time I saw you with Teco. The way you have described today the concrete reality of "ape language research" shows that seemingly small decisions will have consequences for the kind of communication one could arrive at. The "physicality" interacts with the conceptual framing. We have explained how the hypothesis of some "purified" language ended up in the last decades with the promotion of language being deeply "pure" and "perfect." I would take exception with even the possibility of getting absolutely purified language within language (formalized mathematics is not the truth of an idiom). One can create situations where some effects of language will be *neutralized,* but this is not the same as "purifying." Commonalities of reference, of culture, of speech regulations (from grammar to the methods of the disciplines)—all these factors as well as others could neutralize, to a certain extent, the nonalignment of words and concepts. Now, granted, we are not saying that syntax has no function in making language work. Not at all. We are critical of the way idealized grammar has been promoted as *the* decisive element of language.

SUE. Yes.

LAURENT. What we have said so far was also part of our *preparation* for a discussion of the use of human language in chimpanzees and bonobos. Quite logically, we may wish to move on to the next part of our dialogue through the question of syntax, or of morphology and syntax.

SUE. Agreed. The work with Lana, that Duane conducted, was designed from the start to approach the issue of syntax. Duane made the same assumptions the Gardners did: symbolic capacity is there, it is fairly trivial, and to confirm it, we have to evaluate sentences. Duane started with what he called "stock sentences": "PLEASE MACHINE GIVE MNM," "PLEASE TIM MOVE INTO ROOM," etc. As she was using her computer, Lana didn't utter just one word like "APPLE." In order for her to get an apple, she had to press a series of keys and express "PLEASE MACHINE GIVE PIECE OF APPLE." In contrast, she could not say "PLEASE MACHINE GIVE PIECE OF MNM." Lana's entire language and the syntax that she employed were designed by the linguist Ernst von Glasersfeld. There was no linguist involved in what Herb Terrace or the Gardners did: whatever those people thought of as being *syntax* was used with Washoe or Nim through sign language. Von Glasersfeld designed a syntax that superficially resembled English syntax, but was distinct. It was designed to enable von Glasersfeld, as a linguist, to determine whether or not Lana was putting what he assumed would be symbols together, and whether her referential capacity had syntactic components.

In that very well-conducted study, three major obstacles appeared. There was a first problem that was doomed to occur, since Duane had not really addressed the issue of symbolic capacity and whether or not grammar is *inherent* in true symbolic capacity. Because of this oversight, there were bound to be conceptual issues that would pop up, that could not be addressed by the experimental method and that were not answerable within the context of the work. Duane, like the other scientists I mentioned, leapt over the issue of symbolic capacity and its grammatical form, not realizing it ever existed. What Lana did, or did not do, what she could be said to have done or not, was not a realistic discussion because the experiment was not designed to look at it. When people dismissed Lana's behavior along those lines, it was an arbitrary dismissal. I was one of them . . . I now have a very different perspective . . .

The second problem was a consequence of the way the computer had been designed. As never happened before, or ever since, the computer was programmed to collect *every* utterance that Lana made, as well as *every* utterance anyone else would make in the experimental setting. The Gardners could certainly not do that. With Lana, the initial feeling was that if one could record every single utterance, Ernst would be able to do a syntactic analysis to check if Lana were really picking it up. Lana was always using language in a context. Tim Gill, the PhD student who was working with Lana, understood the context as well as Lana did. But the context was not being recorded by the computer. This is what Terry Deacon brings up in *The Symbolic Species*: each symbol is always contextually embedded within a whole other matrix beyond the word syntax, and, in order for this symbol to become referential, it has to be related to this context in precise ways that both the speaker and the listener jointly comprehend and are aware of. Terry's argument is absolutely true. When symbolic behavior arises and becomes embodied in intentional, referential communications between two beings, grammar is already employed. It is impossible to have symbolic referential communication between two living beings that is not grammatical. The grammar is not referring necessarily to other words within a sentence: even a one-word communication, in the right context—with the grammatical links to gestures, glances, objects, and to the previous utterances—is symbolic. Thus, grammar *must* exist. But when von Glasersfeld found truly structured novel utterances with multiple words as recorded by the computer, he did not know what was going on in the context. So, he couldn't make any definitive analysis, which shows the flaws in linguistics, up until that time. Von Glasersfeld was upset he didn't have the context, but he had not told Duane that he needed it, so there had been no effort to record it! Ernst then asked Tim to record the context in addition to what he was saying to Lana. This was really difficult, and, at the time, there was no easy computerized way to do it. Who can record what they are doing without stopping what they are doing in order to record it? . . . Moreover, the context was too broad: should Tim record all the objects that were around him and whether he had moved them? The context is everything, it is everywhere, it has no clear dimension in time. In their interactions, Tim and Lana could be referencing sessions they previously had several days, weeks, or months before. But how can you tell somebody else that you have had this joint memory? And how can you record it in data and state: "I know that Lana remembered *x* when I did *y*"? The issue of how to record the context quickly became overwhelming. In the meantime, von Glasersfeld, working as a linguist, thought: "It should be easy to record the context! Why can't they record it? I just need this one little piece of data about context, and, if I had it, I could tell you whether this utterance was, or was not, a syntactically proper sentence in my von Glasersfeld Yerkish." I only understood much later that the original Yerkish was a really well-designed language. But even linguists must have some knowledge of the context to make decisions, unless they are dealing with high-level human texts where everything is packed into the words. If one is dealing with situations where grammar is not completely packed into the words but ostensibly seems to be there, this is a very different situation. There was so much going on that Tim did not know what Ernst would later need. Without the same operating system in your brain, this system we can call language, there is this "blooming, buzzing confusion" William James talks about.[10] The operating system in your brain is suddenly aligning how you feel, how you perceive, how you think, so we begin attending, thinking, and building our worlds in a relatively similar way.

Let me refer briefly to another experiment, and to when I became painfully aware of this organizing power of language. I attempted to work with Tamuli, who was raised by Matata and was Panbanisha's full sister. Tamuli was clearly brilliant, intelligent, wonderful, but I couldn't align with her linguistically. It wasn't that Tamuli lacked a cognitive ability or an understanding of some kind of language: I was convinced she had some kind of language with Matata, but it wasn't what I or Panbanisha had. Even if this "language" had words and possibly grammar, it was not human. With Tamuli, I could not co-construct how we were looking at and thinking about the world in the same—almost effortless—way I could do it with Panbanisha. I am taking a concrete example. Let us say I wish to conduct a test of theory of mind. To determine if Tamuli and Panbanisha have a "theory of mind," we are going to do one of the Smarties tests: somebody puts little candies in the box and leaves, then somebody else comes in and replaces the candies with something the previous person did not see. Where will the first person look for Smarties? But at the same time we do all this, many other things are going on in the world. From Tamuli's standpoint, these other things are as relevant as the whole spiel with the Smarties. I am using language to draw Panbanisha's attention to the test. But when I begin to try to draw Tamuli's attention—and this is what actually occurred—I cannot tell her what are the components of the test, and to focus on them. Without having common language, there is no way to make this coordination. Coordination is linguistic though it does not have to be in words. It could be with gestures and glances, and it brings our understanding of time and what is important in that time into alignment. Tamuli was not in chaos herself, but, relative to me, she seemed to be.

When context is being linked, even to a single utterance, through a syntactical system that is somehow jointly understood, this is where the "deep structure" lies. It is eliminating plenty of things from the joint consideration of the beings that are in communication, while allowing the interaction to stay on topic, to flow forward across time, to metamorphose in different referential ways. Symbolic processes and grammar cannot be dissociated, as far as I'm concerned. Had von Glasersfeld had video cameras everywhere, and not only on Lana and Tim but on all angles of the lab, his little crystallized boxlike language Yerkish would have shown, unequivocally, that Lana had syntax. Lana had both reference and syntax, though it was a very strange reference and a very odd syntax. It did not have the taste and feel of the language we have, because Lana did not acquire language the way we do—by listening to it. Unlike a human child, she could not sit around and listen to the language being employed around her. She had to communicate either with the machine or with Tim, and to figure out how to do this, she *had* to make mistakes. Lana had to acquire Yerkish through *trials and errors,* by pressing keys and making sentences, and finding out if she were right or wrong in seeing what Tim responded. There was literally no other option available to her. When von Glasersfeld and other linguists began analyzing the data, they immediately assumed that Lana was making errors because she was a chimpanzee, and because human children would not make those kinds of errors. It never occurred to these researchers that a human child raised as Lana had been would have made similar mistakes. So dominant was the idea of species over the idea of rearing as a critical variable that Lana's amazing performance was simply dismissed by Herbert Terrace. In retrospect, one cringes at the hubris and ego of the field's practitioners.

Yerkish had this characteristic: its grammar vaguely sounded like English grammar, but it was different, and only von Glasersfeld really understood it at the time. It took me many years to understand it. Tim fell into a trap, because he knew the symbols and he knew English. He did not know Yerkish the way Lana knew it. There were times Lana used very accurate Yerkish sentences, but Tim told her she was wrong, because he was using English syntax. This created huge difficulties for Lana. Yet, the data Duane collected and the things Lana did really stand on their own. They contain the unassailable proof that Lana had both reference and grammar. It took me two decades to fully come to that conclusion. In the 1970s, I was highly critical of Lana. I am sorry that I was and I wish very much that Lana still had her original system and that she could still be allowed to communicate. I wish I could have the opportunity to work with her today, because now I know what to do. At the time, I didn't. It was a situation that I could not intellectually wrap my arms around. Neither could Duane. Neither could Tim. Von Glasersfeld was trying to help, but he kept screaming "Context! Context!" We didn't know how to give him that. Furthermore, there were so many things that Lana couldn't do, that you would think she would have done, if she had language . . . Lana had to make a sentence and be told it was wrong. That was the only way she could learn. Imagine if this were the only way a human child could learn language—the outcome would be terrible! The child could be very intelligent, and living in a good environment, but if she hears no language around her, and if the only way she is going to speak is to talk to a machine and make errors, and spend a couple of hours a day with a person who corrects her without engaging in a real conversation, well . . . As for oral English, by the time Lana began, she was probably too old to pick up any of it: she didn't seem to demonstrate any comprehension of spoken language. When I look back at what Lana did, I am astonished that, in such conditions, she was able to do so much.

Everything went downhill because Terrace came forward with serialist learning with pigeons. His study of Nim was done to attack the Gardners. And the work Terrace did with pigeons was to attack Lana. Terrace's work never was about language, but on "proving" that others were mistaken. Von Glasersfeld and Duane were, of course, very irritated at Terrace, and, with other colleagues, they tried to do some analyses of Lana's language to show she really had syntax. But here was the problem—and, as you have understood by now, this is the third difficulty I wanted to identify. When you take the whole corpus of Lana's utterances, *because* she *had* to make errors in order to learn anything, one could find lots of errors in the data. So it is easy to argue that Lana was making errors as in serial learning, and that she was conditioned. One needs to make the counterargument that yes, Lana was learning by trial and error. But she learned. So, let us throw all these data out. Let us look precisely at what she did here, within a contextual reference. Let us re-create a description of her world and daily life the best we can, let us analyze utterances within von Glasersfeld's Yerkish grammar per se. If we do this, while maintaining some statistical robustness, then suddenly this other thing that doesn't make sense or that looks like mere trial and error appears in a new light. What Lana is doing is not reducible to language or the absence thereof. She has, all the time, to try to figure out what the computer wants her to do, or what Tim wants her to say, so the errors themselves become revealing. Especially with Tim, because they are probing to get something out of him. She can't listen to Tim, for Tim does not speak to her and she doesn't comprehend spoken English. Errors that are designed to get information out of the world are not errors. They are not grammatical or semantic errors, they are cognitive probes, and Lana's data have never been analyzed from such a perspective. Terrace's critique at the time was sufficiently devastating to stop funding for Lana.[11]

When we ask questions about symbolic and grammatical capacity in apes, I don't think there needs to be any more new data. The data from Lana, from Sherman and Austin, from Kanzi and Panbanisha are certainly sufficient to prove all these capacities. Until there is a certain level of acceptance, it is very difficult to move to the next step as one should do. If one is assuming that these apes have a symbolic layer and grammar, the next step is to begin paying attention to what they say. The step was just taken *immediately* with Washoe and Lucy: "Washoe is saying this! Lucy is saying that!" Then, everybody asked: "Hey! are they really saying this or that?" Even though apes might have such an ability, the way in which the early experimenters took that step was far too early. The field is forever stuck at this level, and trying to justify these initial claims that everybody is now saying were not fully proven in the first place. If we realize they were in fact proven, the next step is to say: "There is some validity to the apes' capacity to say such things. Now I'm interested in *what* they are saying, how it relates to the context—not in proving that the context is layering in the grammar, but in exploring the kind of concepts these primates have of the world, the kind of concepts they have of themselves, and the role language is playing in their mental development."

LAURENT. I believe this is what we have explored and will continue to explore together. Earlier today, we discussed the basic dichotomy used by Descartes—and so many others, both before and after him—dividing each language between *word meaning* and *grammar.* We are saying that we do not have *concepts* and *words* on the one hand (or, maybe, *reference, idea, concept,* and *sign,* which is already more complicated), and a *deep logical structure* on the other hand. We do not believe that the two systems would operate autonomously in language, or that they would simply "merge" in the moment of a speech-act. We do not consider that some species would only have the semantic side, whereas others (or, rather, only ours) would possess the deep structure. Or that some nonhumans would have a version of the two autonomous levels, without the ability to "merge" them. We suggest that, in the communicative process of meaning, a need for organization will arise. It might be the case that this need relies on cognitive structures that are preverbal, to a certain extent, but, even so, the syntax of language cannot be the mere expression, transcription, or translation of nonverbal cognitive structures.

SUE. Language is the actual symbolic layer above life; syntax is a way of relating to concepts through the interface of words.

LAURENT. There, at its core, meaning is a differential operation that is performed in situ by an embodied mind, through all aspects of the interrelations of words.

SUE. Meaning is what is going back and forth between the layers. Meaning goes back and forth between one layer and the layer above it, and the other one above, etc. Then, it is like you mentioned to me the other day about Tom Mitchell's experiment on meaning in the brain: there are differences between categories of words, depending on the role played by reference, whether it is *beautiful, freedom, table,* or *to go.*[12] A word like *the* or *it* has no reference. The only reference *it* has is to something mentioned earlier. Unlike *table,* "it" does not stand for a particular thing, but for a word. You can use "it" a number of times in a reiterated fashion, so that "it" is the same word—or not. Let me ask you: "Is it the it that was the it the other day?" This is a completely interpretable sentence, that you can understand and parse: "Is this 'it' the 'it' we saw the other day?" Sitting here with me in the living room where we were when we famously saw the squirrel we already spoke about, you know exactly what I mean: "Is this squirrel the squirrel we saw poking its little head out of the hole the other day?" You understand the reference because we have a commonality of experience. If you don't have an understanding of the grammar which I am using to put together the words *it, that, the, is, was*—all of which have no referential meaning you can really look up and understand in a dictionary—you do not get *it.* When you begin to see an ape that is either able to put these words together in a structural way, or—quite certainly in Kanzi's and Panbanisha's cases—is able to easily understand those words, not only in one sentence, but also in four or five in a row, you realize that the brain is structuring the input through grammar. You begin to understand that this is what grammar is doing for language. In this sense, it seems to be very disconnected from the symbol. "It" really isn't.

LAURENT. Here, I would like to add a few words on *the* and *a.* The definite and indefinite articles are commonly conceived as being "function words." They seem to carry very little meaning, or possibly none at all. In French or English, one says *the bird flies* or *l'oiseau vole*—or *a bird flies* and *un oiseau vole.* If you omit the article, you seem to produce little to no semantic inflexion: you just become agrammatical (**bird flies*, **oiseau vole*). Or you say something else that will need to be underlined through prosody and/or punctuation—e.g., *Bird flies* or *Oiseau vole,* if you are talking of an agent having such a name or nickname; *Oiseau, vole!* if you give an order to an object or animal named "*Oiseau.*" In some respects, the words *the* and *a* seem to be both grammatically loaded and semantically tenuous (or perhaps even void). This could lead us to differentiate such articles from words like *freedom, apple, to go*—and even from words like *it,* since a pronoun, through anaphor and cataphor, is apt to mean "by proxy," as in your examples. This particularity is so strong that some scholars working in generative linguistics have proposed to consider *the bird* not as a noun phrase (NP), but as a "determiner phrase" (DP), having *the* as the "head" of the phrase. That the determiner, rather than the noun, would be the "head" is a rather amazing suggestion. Plenty of languages, as Latin or Chinese, have neither definite nor indefinite article. Generative grammar could still presuppose that, in such languages, the head of the phrase is the null determiner. Because a universal structure is needed, you end up creating a "ghost article" in languages that do not use one. Beyond Anglocentrism, the main reason behind the theoretical creation of such "determiner phrases" is that they are based on quasi-asemantic "heads," whereas noun and verb phrases necessarily retain some glimpses of meaning in the top branches of the tree. Moreover, the category of DP introduces parallelism for verb phrases, consolidating the purity of syntax. DP is the epitome of "meaningless" and geometrical grammar; thus, it has to be universal. Too bad, it isn't.

Whatever the status of the "determiner phrase" could be in generative grammar, let us consider an idiomatic construction consisting in the omission of the definite article when the noun is taken, for instance, in a general sense. In those cases, one could speak of a null determiner (if need be), but certainly not for languages where the definite article is nowhere to be seen. We rather say *I love music,* which is always confusing for French speakers, who, when they learn English, will tend to say *I love the music.* This rule bearing on the omission of the definite article is nonuniversal, and, in fact, is pretty rare in languages using a determiner. Even "universalists" have to admit it is a specific rule, in English—and not well-ordered "mentalese." But *I love the music* is not so ungrammatical. It is even correct in the right context (in a concert, for instance). There, *the* is close to the demonstrative *this,* which reminds us that a large number of languages use, in fact, this class of deictic pronouns where we'll use the article. Romance languages have either a definite article (preceding the noun), or a suffix: the etymology of those determiners is coming from the system of demonstrative pronouns that Latin was using, in the absence of specific articles.[13] In English, *I love the music* is syntactically correct, whereas **We bring the peace* does not work well. Then, the differences between *I love music* and *I love the music* are first and foremost pragmatic and semantic. If we were to take those two sentences in isolation and in their written form, both should be considered as being equally admissible. But, in context, one option might be "wrong."

More generally, even reputedly asemantic and hypergrammatical words like articles are the outcome of semantic strategies. Syntactic structures differentially encapsulate semantic options. Languages without articles are not *vaguer* than English. They are even able to express that a noun is definite or indefinite, but they do it without relying on dedicated function words. Sometimes, yes, they will create more ambiguity than *we* would expect, but this will be either explained through the whole speech-act, or intellectively exploited, as in classical Chinese poetry. Seeing grammar as a system of organization that is being brought forward by semantic needs largely modifies the role of syntax in the overall conception of language. Groups of speakers gradually solidify rules that are necessary to accomplish semantic communication, as Luc Steels's "talking heads" would confirm.[14] All this naturally bears on ape language research. While Kanzi, Panbanisha, and Nyota understand very well spoken English, "their" Yerkish (which is no longer the crystalline idiom invented by von Glasersfeld) is not English. The fact that they would say "BALL"—rather than "A BALL" or "THE BALL"—does not show that they have no syntax. Unless we consider hundreds of human languages to have no grammar.

SUE. That's right.

LAURENT. Of course, for apes to use human verbal language in a semantic way, they need to have something to say. They need to try to communicate. Here, it is time, I believe, to examine Michael Tomasello's very weird position. Tomasello acknowledges that the beings you call "languaged apes," and that he interestingly names "linguistic apes," use something like human language. It is not full-blown; still, it is there. But whatever "linguistic" aptitudes Kanzi could have, he does not use them to *communicate.* Only humans do communicate, Tomasello says. What a bizarre thesis!

On one level, Tomasello is simply radicalizing the philosophical gesture consisting in excluding animals from the orb of language. In his *Politics,* Aristotle famously considers that nonhuman animals use their *voice (phōnē)* as "a sign for what is painful and agreeable" to them, and that "nature" grants them the ability to "signify to each other" the content of what they "feel."[15] The use of *phōnē* by Aristotle is often eschewed, but it is remarkable, for, in Greek, the term refers to voice, to an articulate sound, to the cry of an animal—but also to some kind of speech. Since Aristotle concurrently uses the words *signs* and *signify,* he is clearly granting nonhuman animals the ability to have a communication system relating to internal states. There is no doubt that Aristotle develops an anthropocentric doctrine and that he sees *logos*—organized and rational speech—as being a human prerogative. Still, he is far more subtle and complex as a thinker than the tradition usually assumes. Furthermore, he devoted many years of his life to studying animals, which sets him apart from almost *all* other philosophers. Clearly, to him, the human specificity of language cannot conceal the existence of animal communication. Some two millennia later, Descartes, in his letter to the Marquis of Newcastle, reiterates the Aristotelian position. He certainly rigidifies the exclusivity of human language. Descartes tends to take the term *sign* as a synonym for *words* and *speech* (both being implied by the French *parole*), so animal "signification" is no longer possible. He concedes that nonhuman animals are instinctually capable of "expressing their passions"[16] (namely, here again, pleasure and pain), whereas humans are rational and also have a system of signs—language—to convey their "thoughts." When Tomasello is denying the communicational use of words in the languaged apes, he just structurally repeats, in (supposedly) his capacity as an experimental psychologist, what has been argued by prominent philosophers from the past: he claims to have found an attribute of language that is specific to humans and unknown to animals. But in asserting that this attribute is *communication,* Tomasello is much more anthropocentric than Aristotle or even Descartes, for he posits that the potentially accurate and adequate use of human words by apes does not matter, since all communicational intent is absent.

This is the first gigantic problem with Tomasello's thesis. Admitting the existence of "linguistic apes" on the one hand and rejecting their ability to communicate on the other hand is, at best, a contradiction in terms. At worst, it is absurd. It is conceptually inept to keep the philosophical gesture of exclusion—that was based on the human uniqueness of having some degree of language—while conceding that Kanzi and Panbanisha may use words properly. Any semantic transmission, if it makes some sense or meaning that is "common" to the agents, merits being called communication. It is ridiculous to excise any verbal "goal-oriented command" from this horizon. Even vervet monkeys' alarm calls have a *communicational* function. Why would they even arise if they had none? Now, are those calls like human language? Not so much—but this has nothing do with the function of communication.

I know that Tomasello, in order to press his point, needs to reconfigure the word *communication,* which he does by attaching to the term a sense of cooperation and of "gratuitous" description. Only humans would use language to show their environment, to comment on it, to express their desires, hopes, and fears. The apes with some command of human language would still be limited by a kind of self-centered usage and would never share with others their thoughts about what is going on. They would never even point at things for the sake of it—whereas human children do this routinely, at an early age, through gestures and, later, through words. We are now arriving at the second problem. What Tomasello says of the usage of language by the bonobos you raised is simply, and overwhelmingly, *false*. It is a very empirical fact that I could experience during *all* the visits I paid to the apes at the lab, even during the very brief, last, one in 2014 that occurred after your removal from the facility and the deliberate effacement of language immersion. I want to be specific here. I have seen, repeatedly, that the bonobos did use their keyboards to ask humans to bring them their objects—especially when they were deprived of them—to deliver food—through precise lists—or to GO OUTDOORS. But, first of all, humans also use language to conduce similar "commands," and it seems very far-fetched to consider that, in such instances, we would no longer "communicate." Especially when there is mention of a *favorite* food or object. We also discussed before that the apes tend to be more "goal-directed" in their interactions with people they don't know or dislike. Secondly, the languaged apes do describe their reality through gestures and signs. They do it all the time.

In the hundreds of hours of video recording, one has plenty of evidence of communicational intent. But I'd just like to mention a few of my own, limited, experiences. To me, the "SAME SHIRT" exchange with Nyota was very striking, and completely clear-cut. Furthermore, Nyota was the one initiating the interaction, and there was no direct demand: a typical example of information sharing, of coelaborating meaning through a speech-act, of "gratuitous" and "collaborative" description. Another case: in the fall of 2011, the day after I had my flu shot, Panbanisha, who had been told that I now had all the necessary vaccines, asked me "SHOT HURT," to which I responded by the negative. Then she said she'd like to "SEE" my shoulder (she showed the relevant body part). As for Teco, when he was sixteen months old, he took me, not by the hand but by the finger (he was very small!), on the very first day of my visit. We went together to what used to be his bedroom, and he pointed at his bed, glancing at me to be sure that I was following him. Then, he took several of his books and toys and gave them to me, one by one. Despite the banal accusations of anthropomorphism that could be thrown at me, I deem it intellectually impossible to consider that the function of such actions is noncommunicational and entirely dissimilar from the exact same series of actions that a human child or infant would do—such actions and gestures being systematically expanded and glossed over with the utmost generosity by Tomasello in his *Origins of Human Communication.*[17]

SUE. Thank you for saying this, Laurent. You know, when I was a graduate student, I encountered two different kinds of language worlds. One was a group of chimpanzees that included Booee, Bruno, Cindy, Thelma, sometimes Kiko and others, that had either been wild caught or born in captivity but were pretty much peer reared on an island. Another group was made up of chimpanzees like Lucy and Solomon: they had been taken from the day of birth and placed in a human home. The family living with a chimpanzee was asked to raise it as though it were their child. Those two groups of chimpanzees had been established before Roger Fouts came. Roger Fouts had helped the Gardners with Washoe, and he came to Oklahoma with the ostensive goal of teaching language to these different groups of apes—the ones that had been reared by their peers on an island as well as the ones that had been reared in human homes. A great effort was made to have somebody staying with the human-raised chimpanzees all day long. By the time I arrived there, Lucy was around six, so there were periods of the day when she would just stay in a cage, and her family didn't feel this was a good solution for her. Otherwise, Lucy had been constantly raised with, and by, them. Lucy had slept with them at night, she had all her meals with them, and they talked to her. She had a human sibling that was six or seven years older. Roger Fouts would spend maybe an hour a day with Lucy and would hold up objects, and mold Lucy's hand into the sign for that object. This was how he would "teach" signs to Lucy. The family didn't really know sign language, and they already had a relationship with Lucy that wasn't based on such signs.

The apes from the island group had had a wide variety of previous experiences which were important, but at the time, they were all spending each day together, in relative isolation. Roger would regularly take with him, in a boat, the chimpanzees living on the island. Off the island, the apes would stay in a cage, and students such as myself would go into the cage and hold up objects and teach them signs. Then, the apes would go back to the island. Whatever language they had, it could only consist in what they received in the training sessions. I knew this was different from the training method of the Gardners, because they tried to keep somebody conversing with Washoe all day long. Overall, the apes were not really living in languaged worlds. Roger was trying to draw some conclusion about what the chimpanzees could do, based on what he had taught them, but apart from what they had learned in sessions and the tests, sign language was not in their life. This was also true of Lana: Tim was only working with her a few hours a day, and when he left, she was just with the computer. This was also true of the work by David Premack.

I decided that, if I were going to compare apes and children in any way whatsoever, I had to try to follow the paradigm of the languaged world that the Hayes had used and the Gardners had tried to use. I had to give the chimpanzees an environment where people would constantly be using language with them. This is much more difficult than just going in for an hour a day then testing for "language." When you give apes some access to this world, you can't be training them to do signs all day long; you must have a world that is interesting. I first tried, as much as I could, to accomplish this with Sherman and Austin. I was able to expand this world a great deal with Kanzi, because, thanks to the forest, I could emulate his natural environment to a certain degree by traveling through the woods with him. When you say that the bonobos were using language all the time, yes, before Jared Taglialatela and Bill Hopkins took over in 2013, they were . . .

LAURENT. The bonobos also used language to comment on what was around them.

SUE. Well, yes!

LAURENT. Tomasello's point is that, even if they were to use some language, it would only be to *get* things.

SUE. That is his point, but it is just . . . untrue. I initially reported on this issue with Kanzi. I had large bodies of data where I reported all of Kanzi's utterances: a large chunk is not about getting a piece of food or making a request. And, certainly, there is always this desire to show their world to the visitors. In Atlanta, Kanzi liked showing his whole forest. Teco wasn't allowed to have this, so he showed you his world. Teco is not going to do the same with me, because I am in his world.

LAURENT. Of course. Teco did it only once with me, on the first day of my stay. He did not repeat the behavior on any of the following days.

SUE. Yes. In life, with Teco or Kanzi, many things are happening only once because it does not make sense to repeat those things again—unless they are a ritual. If so, Tomasello could argue that a ritualized behavior is non-linguistic in a sense, which is one of the arguments he has made. Apart from rituals, you have to be prepared to grab everything and inscribe it in the proper context of its first-time behavior. This is difficult. This is not what experimental psychologists routinely do.

Most of those apes' behavior has nothing to do with getting food. This was even more the case in Georgia than in my time in Iowa, because, in Atlanta, the food was distributed throughout the environment, throughout the whole forest. Kanzi would not *need* to talk about the food, he would just need to get it, or fix it, or eat it. Thus, conversations could deal with everything else. Once we moved to Des Moines, I wasn't permitted to re-create the world in the way I had anticipated. Food was only in the kitchen, where the bonobos could not go. It wasn't distributed where the apes were, whereas even Lana had her own access to dispensers. The only way Kanzi and others could get food was to ask for it. Then they wanted to store it and be able to get it when they so wished. By simply removing the food from the apes' reach, a situation was created where they would *have* to ask you to bring it to them. I didn't want Kanzi to ask people for peanuts or onions, or even for their permission to open the refrigerator. I wanted Kanzi to access his world the way he would if he were a free ape. Basically, how often the apes use language to ask for food, or whether they do it, is not a function of their cognitive capacity. It is a function of how you structure the world around them. Nothing more, nothing less.

LAURENT. The same goes for commenting on the world. If there is nothing to describe, the apes will describe nothing. If a bonobo primarily lives in an enclosure with bare concrete floor, a few hoses and platforms for climbing, a blanket, and maybe a ball, but no musical instrument, no book, no magazine, no toy, no computer, no TV, and very few people to see, why would this ape, even if it possessed linguistic capacities, describe its environment? Beyond the theoretical critique I would level against Tomasello's statements on noncommunication, this other consideration, that is much more matter of fact, also holds.

SUE. I have ignored what Tomasello wrote on the topic, in part because Patricia Greenfield, in a very eloquent set of data involving Sherman and Austin, already made this case.[18] Tomasello is perfectly capable of reading that paper and understanding that the point was made, even before Kanzi. I know Tomasello personally. He has been to my lab, and I know what he saw there. If he is making this critique of my work, ignoring the data that I know he knows, what should I say? I might say something to the rest of the world that would seem to negate what Mike argues, but this would be a kind of game. When people who are supposed to be your peers in the field ignore what you know they know, and what they know you know, all this is no longer a scientific issue, it becomes a political one.

LAURENT. This additionally happens in a field that does not have so many qualified practitioners. Which makes it easier to sell falsities.

SUE. When Mike Tomasello is willing to say—which he does—that we have to take the abilities of linguistic apes seriously . . .

LAURENT. Which is already significant . . .

SUE. Sure! Mike has been in my lab enough and he has worked enough with the same chimpanzees I have worked with to be able to make this statement. He knows this to be true. So, when he does this, he also tries to come up with something that fits his own schema—even if he knows better. Mike Tomasello was chosen for his position at the Max Planck Institute because they wanted to investigate the phenomenon of language. They invited Mike: they knew he was working with me, therefore he was a good choice. Mike came to me and said, "They have invited me, they want me to do language research, I don't know if I can do it, it's a big job, it's a big change, what do you think?" I said, "I think it is a great opportunity for you." I was hoping he would do some good language work, because he certainly had the knowledge and capability. I agree with the people who chose him: he was an excellent candidate. Then Mike, before he ever went to Germany, told me: "I have thought about it more than once, Sue, and I have decided I would not try to raise any chimpanzees myself in a language world." He made that decision. There were different reasons he gave me: that it was an ethical decision, or too difficult to do with the graduate students that he would have and the structure he would be in. At any rate, once he made that decision, then he committed himself to work with the animals he was being provided, with primates reared in a zoo context, which have very different backgrounds. By the way, you will notice that he never gives any information about the background of any ape participating in his studies, even though he was trained, first and foremost, as a developmental psychologist and is well aware of the plasticity of the brain.

Tomasello committed himself to come up with a framework that would explain what he saw in these different apes. He pushed linguistic apes off into some domain that he says he will not study or understand. Then, within his particular framework, he is bound to come up with an answer that will satisfy questions about "primates" in general. Tomasello has published a lot and he created a huge cadre of students who are pushing this very German, Max Planck Institute, view of innateness, instinct, and "species-specific" skills—a view that is also echoed by others authors like Josep Call, or Brian Hare. I have generally avoided critiquing Tomasello's research, because, by doing it, I incorporate personal knowledge. I have not done an experiment to "disprove" what Mike says, because my data are already out here. If Mike is affirming it is all right to ignore the data, it all becomes too personal at that level, although it is the truth.[19]

LAURENT. Maybe we will end up here our response to Tomasello's thesis about the uniqueness of human communication.

SUE. Yes.

LAURENT. Sue, this was a long dialogue on the theory and practice of human language, as it is being transferred to apes . . . But before really ending our conversation, and since we have touched on several critiques that were voiced against your experiments, I wanted to ask you a much more circumscribed question. It might seem to be tied to most of the issues we faced today, including the limits of what could be said and the importance of meaning and grammar. In what I have been able to read from you, and what I have seen myself, Kanzi, Panbanisha, or Nyota rarely combine more than three signs in one continuous utterance. My question is simple: Why do we have relatively short sentences overall, instead of having much longer ones? One could think that the control over syntax is limiting the length of the sentences. Or that bonobos, even when they do communicate through Yerkish, are unable to come up with strings of words a "normal," human, four-year-old child would master.[20]

SUE. Okay. First of all, Lana did not have that limit.

LAURENT. Yes.

SUE. Lana first went up to five symbols. When the limit was made seven, then nine, the experimenters themselves had become very confused about Yerkish syntax. Tim was already nearing his PhD and was ready to leave. Nobody could really replace him because nobody had the understanding of Lana that he had. Because talking to Lana wasn't just talking to her: you had to know how to do it. I tried to talk to Lana. Sometimes I could, and sometimes I couldn't. If I had gone back and really trained myself, I might have been able to talk to her. But I did not put forth this effort. By the time utterances were seven to nine symbols, I noticed that Tim had begun to cue Lana with his eyes. Moreover, although Tim could ask Lana to execute things that she would do, I didn't see any general comprehension of language within her. It all seemed very specific. Back then, I didn't understand how specific it was to Yerkish grammar. The way Teco could understand you when you interacted with him and he was less than two years old, Lana could not do that. She could not do it, even today. If you just look at the list of utterances, Lana easily did use five symbols. She started using long chains. Within the proper conditions, she would have mastered seven- or nine-symbol strings.

When, after the Lana experiment, I started with Sherman and Austin, I really wanted to avoid the cuing. I was also concerned that a chain would make things too complex to analyze. In addition, when I began the Sherman and Austin work, other colleagues had worked with those chimpanzees for two years, and no learning of even isolated words had occurred. Duane had already made the choice not to replicate the Yerkish syntax with Sherman and Austin. He had decided to start with single words and ask questions about single words. I was brought into this situation. I had seen the trouble Lana was having with longer sentences, and I didn't know how to solve the problem, so I tried to avoid it by using shorter ones. Once Sherman and Austin became referential, they could combine and make words with two or three symbols. They had implicit syntax in there, but it was much more variable than Lana's, since they were not trained in Yerkish. But, on their keyboard, one could find none of the function words that were on Lana's computer. Originally, A, THERE, THE, IT, AND, ME, etc. were not on Kanzi's or Panbanisha's keyboard either. The words you would typically use to make longer sentences (like AND, WITH, THAT, HOW) were simply not available in the language used by the apes. We did not know how to teach such functions, aside from embedding them in a sentence requiring the apes to produce this type of sentence. One can examine how "THE" is employed in sentences. This had been clearly done with Lana—but the rest of the world had concluded that, Lana lacking syntax, she could not understand words like "THE," even though she used them appropriately. This was a "Catch-22" situation.

When I realized that the bonobos were understanding spoken English, I saw that they were understanding those function words. I also grasped that I was saying those words thousands of times more each day than I was saying the words that were actually on their keyboard. So I introduced the relevant lexigrams on the panels. This happened very late in Kanzi's and Panbanisha's lives. But after about two years, Kanzi and Panbanisha did start to use them. It would be infrequent, and it wasn't in the majority of their sentences. Still, when they used them, it was correct. It would usually be a part of a sentence where the rest would be inferred. So the utterance might be only three words long . . . I began to feel the bonobos knew these words at a level Lana never reached, because they understand them within English grammar. Lana only understood them in Yerkish. There was another obstacle, however: I tried to get the caretakers to use those words in longer sentences, but I never succeeded in getting anybody else to learn how to make long sentences fast. None of the humans were making utterances longer than two or three symbols. People would say it was taking too long and it was too cumbersome to look for those additional words. Even me . . . While it never took me much time to learn the concrete nouns or the verbs, I always felt I couldn't point right at the function words. I always had to look for them on the keyboard panels. This is very frustrating when you want to get the communication out. There is a time dimension at work, so you end up favoring an easier solution.

LAURENT. You prefer saying "I WANT BANANA" to "I WANT A BANANA . . ."

SUE. Well, I just prefer saying "BANANA! . . ." These long sentences are only easy to do if you sign or speak. They are not easy to do with a keyboard. If you are going to do them on a keyboard, you need to be in a formal situation, like Lana was. You need to practice like you would do with a typewriter. You need to have all the people outside of the enclosure also modeling the use, with their keyboard. When I began to understand all this in Iowa, I had no position of authority to tell the staff how I wanted them to use the keyboard. In brief, to recapitulate my answer to your question: the length of utterances does not reflect a limit of the cognition of Kanzi, Panbanisha, or Lana. Apes might well have a limit—a limit that restricts them in ways that we are not restricted. But, if so, this has not yet been demonstrated.

On Free Will

The May 2015 federal hearings were just over. Giving sworn testimony in an Iowa courtroom about the mental capacities of apes was a rather eerie and unexpected experience for me. The legal team of Bonobo Hope and the different people supporting Sue's cause had done their best. We now had to wait for the decision of the judge, who had been asked to consider if the new people overseeing the life of the bonobos were indeed pursuing the initial "research trajectory." Because the Iowa laboratory was so costly to operate, in serious risk of flooding, and with no access to a safe, large outdoors area, we were also suggesting the apes be moved to a new, and more appropriate, facility that Ryan Sheldon, a tech entrepreneur who had worked as a "computer wizard" with Sue in Atlanta, had offered to build on the land he had acquired in his city (rather, village) of birth, Osceola, Missouri. But truth be told, by then, we were already well aware of the fact that no verdict would put an immediate end to an inextricable situation.

The exchange I had with Sue the day after the trial first touched on what she would do with the new lab if the federal decision were favorable to a move of the bonobos to a site whose configuration would be much closer to what they used to know in Georgia. A very different set of questions would appear, if, in fact, Kanzi and his family had to stay in Des Moines. Sue had to wonder what would be best for the apes, and, contrary to the other side, she was of the mind that the bonobos themselves ought to be consulted about their fate. "Inevitably, then, we should consider the concept of free will," Sue said. I agreed, quite voluntarily I thought—or maybe under the pressure of the events we were facing.

LAURENT. Since you want us to discuss free will, I would like to make a few introductory remarks about the expression itself. I am first delivering the following statement: I do not personally believe in absolute freedom. Of course, freedom is a politically necessary concept, though there should be little doubt that it is a structural illusion, or a promise that will never be fulfilled. No social group will grant total freedom to its members; no social group will be freed from death. I am not saying that we should try to seek happiness in slavery: despite my doubts on *liberty,* I do believe—with intense conviction—in the value of *liberation.* Thus, at the individual level, my free will could only consist in the process of its liberation. And we run into the difficulties of what "*will*" could mean. Incidentally, the Romance languages have generally adapted the Latin phrase and rather speak of a "faculty of deliberation" *(liberum arbitrium)*: the term of *will* is absent. Then, we can wonder if we do have a will that would be ours. I assume we'll refer to Schopenhauer in the course of the dialogue. He clearly considers the individual will to be commanded by the general Will of Nature. In this case, we do not even find liberation through the action of our "will," for the latter is precisely the conduit for the realization of superior determination. In this scenario, *free will,* alas, is just a "representation," just another fantasy that we have and makes of us the puppets of Nature.

SUE. In the field of science, we encounter the idea of determinism.

LAURENT. Yes.

SUE. There are laws that govern the universe, and they are producing all these phenomena. In the end, if we could understand the whole system of laws, we could clearly see that everything is determined.

LAURENT. It goes without saying that scientific determinism is supposed to be very distinct from the brands of religious fatalism one finds in Calvinism or some forms of Islam *(inshallah)*. Though I can certainly see the divergence, I also notice the deep resemblance between the two approaches. I am tempted to argue that "the laws of physics" play a discursive role that is comparable to the one played by the gods. I am not speaking about equations, about the experimental and mathematical description of events in the universe. I am thinking, for instance, of the widespread personification induced by phrases like "the laws of physics dictate."[1] I see a sort of conceptual transfer leading from the action of God's will to "the will of Nature" or to "the laws of physics."

SUE. We were speaking the other day about emergent phenomena. Something blows up and, suddenly, there is life; or it blows up and, suddenly, there is human consciousness. This is basically saying that it is an accident—but within a greater law of how things work that really falls into scientific determinism. It seems that either you find scientific determinism and the forces of nature, which are really representations of the gods—or a God that has created all of this and whose design equally determines reality. The choice is between a determinism without a demiurge, filled with chance, accidents, and time—and a determinism with a creator whose will it is that these things happen. Let us call those Determinism 1 and Determinism 2, okay?

LAURENT. Okay. I just want to note that Schopenhauer, while he speaks of Nature, is rather on the side of Determinism 1. More important, we can find alternative deterministic approaches that include godly agents, without having a creator, or where creation is a minor consideration. Determinism could be based on the will of the gods even though their will is much more accidental. Outside of monotheism, one does not *need* a huge design, though there *may* be one. This is a sort of local determinism, with a plurality of gods that do not have to be the creators of the world.

SUE. The world must still be created.

LAURENT. It depends . . .

SUE. You are saying that, in some religions, the world just always existed.

LAURENT. Yes. Or the gods appeared therein. The gods may have been created in some sense, in which case one goes one step further: how is it possible for the gods to be created?

SUE. All right.

LAURENT. We have the possibility for a third kind of determinism, and it is more "local" than the other two. The wind, or Zeus, or Papa Loko at one point makes a decision that could be debated by others—by other gods, or by superior entities, but also by other human beings, and possibly negotiated in a dialogue. Still today within polytheism, you are asking one god—Ganesh, for instance—to help you in this endeavor, anticipating that another god might be against it. In polytheism, there is the possibility for local determination (without free will) to be granted to human agents. The will of God makes things more monolithic—also requiring the whole universe to be represented as an "act of God."

SUE. We would have Determinism 1, Determinism 2, and Determinism 3. In this latter type, free will exists but it belongs to the gods, and man has to address the gods because he doesn't really have free will.

LAURENT. Humans could also have some level of free will, they are not always blind—but the *consequences* of their deliberate actions will ultimately conform with a superior design. I am going to explain this in a second, but I also wanted to stress that one could think of a kind of nonscientific (or pseudoscientific) Determinism 1. You know that the most ancient written records in China are bone oracles. It has even been argued that writing had been invented in China for the sole purpose of recording the predictions made to the king by reading geometrical shapes that are extant in the natural world, like the cracks of a tortoise shell.[2] The *I Ching* is a complex system of prediction that inherits those early practices. You find similar modes of divination all over the world, like the *fâ* in Benin, the diverse sorts of astrology, and all the "mancies" that require no visionary power or prophetic inspiration. In the absence of God or of the gods, and without having an epistemological theory that one would today call scientific, it is possible to claim concurrently that the future is already written and that the rules for understanding predetermination exist. From there, we could have all sorts of combinations. Isaac Newton is a good example of a proponent of scientific Determinism 1 who also believed in sacred geometry, a creed deriving from nonscientific Determinism 1 that was affixed to some Christian Determinism 2.

Now, back to Determinism 3. Determinism is the stuff Greek tragedies are made of. In this theatrical world, it sometimes seems that free will is just an overall illusion: everything mortals do is a decree of the Olympians. This is often what the chorus expresses—a polytheistic version of Determinism 2. Sometimes, heroes decide, out of their free will, to commit an act that the gods may disagree with: this is how entire families are cursed—on the basis of one initial barbarous deed—and become wholly determined in the future. Sometimes, a hero is being possessed: in an act of frenzy that has been caused by a god or a goddess, he is no longer himself and he does something horrendous that he would never accomplish if he were in his right mind. Heracles kills his wife and children in a moment of madness that Hera elicited. Ajax's reason is clouded by Athena, and he slaughters a flock of sheep. There, the goddesses have their revenge, and the free will of the characters has just been locally annulled. Or, at the very least, this could open up a discussion, onstage or beyond, about the extent of fate. Sometimes, the heroes have some appearance of free will. In a given situation, they really decide what they want. But, later on, it turns out that the *consequences* of such human decisions correspond to the predetermined intentions of a god. Oedipus is a good example. The initial oracle that was given at his birth was accurate: Oedipus will kill his father and marry his mother. However, this does not automatically imply that the decision of his parents to send him away as a baby was *directly* caused by the gods. The idea could be that, whatever route one chooses, the *final* outcome is determined. This does not imply that there is just one route, or that each situation is amenable to direct causal explanation. In the Homeric epos, and in addition to this bounded free will among humans, the gods themselves are debating what they do, and they could contradict (or counter) each other. All this, really, is a feature of polytheism in general. There are discussions, in Greek philosophy and theology, about the determination of the behavior of the Olympians. When Zeus is in love, does it mean that he no longer has complete free will? Does it mean that Eros the god of love is the one determining things, even at the level of Zeus?

SUE. If there is one god that can influence all the other gods through love, this is close to monotheism.

LAURENT. In the same way, philosophically, Plato's Socrates refers to "the god," which tacitly posits a godly principle that would transcend the specifics of individual gods.

SUE. Even though we have three levels of determinism, at some point the third level merges into the second. I am not trying to talk about a timeline here, I am just trying to talk about categories.

LAURENT. In order for Determinism 3 to become Determinism 1 or 2, one needs some general determinism ruling over local determinations. But even love . . . Is love really what shapes the entirety of the will of the gods? No: love, from time to time, is only able to be more powerful than the individual will of Zeus himself. Determinism 3 is more discontinuous and circumscribed than the other two forms. When you go from 3 to 1 or 2, you are simplifying, you are intentionally unifying what used to be both plural and in pieces.

SUE. In Determinism 3, not everything is determined, either in the lives of the gods or in the lives of the humans. Is this what you are saying?

LAURENT. There are determinations without unified, total determinism, nor pure contingency. It is some disrupted determinism.

SUE. If we're going to be discussing free will, there is no place for it in Determinism 1 and Determinism 2. Determinism 3 allots some categories of free will. But Determinism 3, as I understand it, is not really part of our modern worldview.

LAURENT. In Europe, or in the United States, it is not so much a part of it. Or it now plays a marginal role. Many people in the West, however, practically believe in local determinations, without ratifying overall chaos or general determinism. They do not need the gods of Hinduism or the Olympians to believe in this.

SUE. But most of the world's religions are monotheistic.

LAURENT. It could be the case that most of the religious people on the globe now abide by monotheism, but there are many beliefs, and religions, outside that frame.

SUE. Okay. Could we agree that, in *this* part of the world, the debate is more centered on Determinism 1, and, to a lesser extent, on Determinism 2?

LAURENT. We sure could.

SUE. Take this recent book by Sam Harris on *Free Will.* In his essay, Harris is arguing from the strict behavioristic view which one is exposed to in psychology when such questions are raised: the view that everything is developmental and historical. Harris starts out with an evocation of the 2007 murders in Cheshire, Connecticut. Two guys, named Hayes and Komisarjevsky, enter a private home, and they just decide to rape, harm, and finally burn two teenage girls along with their mother. Harris details all the intimate intricacies of the crime: they are horrific. Also, it is as if the crime were committed without much intent. The perpetrators just do things. They have very little awareness of the horror they inflict upon their victims. It is as if they were kids pretending they were somebody else.

After this description, Harris offers the following explanation: all this comes out from the backgrounds of these two individuals; they are conditioned by similar events that happened to them; they did not really have any upbringing that could lead them to understand they could control their own behaviors. In brief, these individuals are the victims of their genetic makeup and of circumstances. They couldn't not do what they did. Harris is presenting this terrible crime to say: even though these people are going to be held responsible and accountable in our society and we know they are guilty, I, as a scientist—especially as a behavioral neuroscientist—gain nothing by calling them guilty. I gain by understanding their background and that they could not have done otherwise than they had done, then I gain even more by trying to build a society that is different. But, of course, my background has led me to write this book, so whatever I am doing has no greater degree of free will than the two people who committed the Cheshire murders. If you take this logic to its extreme, which is what Harris intends to do in his book, you have to come up with some other explanation for even constructing a moral society. The only other realistic explanation being some brand of Darwinism. There is a need for greater good, but it is deterministic and unconscious.

Let us posit now that you have read enough science and thought about those philosophical issues long enough. You also studied theology enough, and you do not feel comfortable saying "There is just some God I can't see and don't understand, but he created all of this for a purpose I cannot grasp." Thus, Determinism 2 seems so irrational and untestable that you reject it. On the other hand, as you consider the scientific explanation and buy the argument made by Harris, you could ask: Why bother with it, since it is all determined anyway? Even your potential interest in Determinism 1 is determined. All these accidental events put you in this position. This idea might be fine as long as everything is going well in your life, but let us say things happen to you like they happen in a Greek tragedy. You don't really like it! It becomes a little uncomfortable thinking it is all determined anyway. Moreover, you cannot really step into the role of the Greek hero that may ask the gods to make a different decision at the end.

LAURENT. If I may interrupt you, in some areas of Greek philosophy there was something called *argos logos,* which could be translated as "lazy argument." This is saying that if everything is completely determined, then we can stop doing anything. In fact, if things are purely random, the result is pretty much identical. We should just wait for things to happen, because we have no grasp on them.

SUE. That's right!

LAURENT. Why should I even bother eating, because if I have to die, I have to die, and there is nothing I can do against it.

SUE. That's right! But the answer lies in the existence of this desire, of this driving force, that is either coming from the gods, or from the laws of nature, or from whatever other instance that is propelling you forward, beyond your free will and beyond your control. This leads to an interesting question that I find not sufficiently dealt with. Let us just speak of a *perception* of free will. Let us take the explanation that was given to me in the sixth grade when

I was a Presbyterian: determination and predestination are these two trajectories and they meet above the clouds. And above the clouds somehow . . .

LAURENT. They are the same.

SUE. We don't know if they are, but, at any rate, the answer will be there, and you, as a human being, are not going to have the answer in this lifetime. If this is the solution, it does not tell you what to do in the meanwhile. It gives you no clue. In particular, when things don't go your way, you resent it, despite your feeling of being determined. We face the same problem with determinism, whatever language, or whatever symbol, we want to put it into. The problem is the same in Determinism 1 and Determinism 2: either God decreed it, or something else caused it. Determinism 3 is an interesting option . . .

LAURENT. With Determinism 3, this is no longer "Why is God angry at me? Why did he forsake me?" Because, with a multitude of gods, we could find one that may have good reasons to be irritated by our behavior and abandon us. There is a particular god that watches over commerce, this one who is in charge of love, or that one who cares about the way we treat our parents.

SUE. In this scenario, we have an idealized world that is allowing humans to develop some degree of free will, and some degree of supplication to the gods, who apparently have at least some degree of free will themselves.

LAURENT. Because of monotheism, we have lost from sight the intellectual beauty of this particular "third" viewpoint that is certainly much more powerful as we deal with issues such as free will.

SUE. By constructing this kind of reality, we give humans a modicum of free will within an organized universe.

LAURENT. Yes.

SUE. In our scientifically driven determinism, we tend to think of free will as an illusion. Just like the self, it is a perception.

LAURENT. Exactly.

SUE. Still, this is a very prevalent perception. Almost every human being, except a zombie or a hypnotized person, retains it. Why, in the face of an overwhelming intellectual argument about determinism, would any well-trained scientist need to retain that perception? Yet, ubiquitously, it is retained.

LAURENT. Schopenhauer, who was calling such perception a "representation," was one of the most prominent intellectuals in nineteenth-century Europe to introduce what he understood of Zen. The end of *The World as Will and Representation* is a plea for impassibility. Once we fully understand that we are being determined by another force, we need to let go, to abandon our illusion of free will, and become indifferent. This is a view that Schopenhauer found in Zen, among yogis in India, in Greek stoicism, in Catholic quietism, etc. The only thing for us to do is to reach a state where nothing matters anymore. This does not seem to be what most scientists do. As for Schopenhauer, he continued to write books afterward . . . He was not completely in tune with his own philosophical conclusion—but this does not say he was conceptually wrong. In fact, if we abide by absolute Determinism 1, the soundest thing to do is to absent ourselves from any temptation that our "representations" could present to us. I guess the American, pragmatic conclusion would be to make money and enjoy our determination as long as we can.

SUE. Scientists and philosophers may have a problem with absenting themselves from their perceptions, but yogis or monks in certain theological traditions are able to reach this state of impassibility. You need to give up this idea of yourself as a causal agent and realize that you do not have free will. You need to give it up in order to reach a higher plane of understanding or being. In those traditions, as you give this up and realize you must think in larger terms, you are nonetheless assuming that you can move to a higher level of existence, or that a monotheistic entity is governing the universe. You are not making the decision that a scientist makes, and you do not propose that determinism is just a big pile of accidents gone right.

If we ask scientists to go from being what they are to becoming high-level Buddhists, this will be a big jump, and they will need to make a long trek from this perception of self to the perception of self as nonself. The perception of self and the perception of free will are very tightly linked. They are not concrete entities. The self-reflexive self perceives itself to have free will. It is impossible to me to think of how humanity could construct a moral society if it did not have a perception of self and of free will. Even the perpetrators of those terrible crimes have existed and still exist in a society which perceives them as having some modicum of free will. Even though a psychologist may explain all this away, society is not going to buy the argument. It is going to hold these people responsible. Could we design a society in which everybody would have so much training in psychology that we'd find a way of dealing with the people who are nonfunctional without keeping any assumption on free will or morality? Is it even possible to have anything we could call a human society without even retaining the slightest free will at any level?

The Determinism 1 that is coming out of science has led to the modern world we live in. If we lived in Determinism 2 or in pure Indeterminism, if the cause to each thing or event were ineffable or "God did it," there would be no motivation for science. Particularly in the behavioral sciences, one posits that every behavior has a cause, that cause has another cause, and so on—then one comes to this conclusion that free will is not extant in reality. If the self and free will, that, in my view, are the two most important perceptions in humanness (the third one being meaning), are claimed to have no existence, can you even have a human society that would be able and motivated to construct any science?

LAURENT. Questions like this (or the one underlying the "lazy argument") enfeeble Determinism 1. Furthermore, hard-line determinism is almost always adopted *with exceptions.* This immediately weakens (possibly explodes) determinism *as such.* Harris's book, in this regard, is a rather sophomoric effort to preserve the purity of a theoretical position that is untenable, either theoretically or practically, or both. I am not a major admirer of Daniel Dennett's, who once spoke of Kanzi out of sheer ignorance,[3] but his rejoinder to Harris is certainly worth reading.[4] What analytic philosophy, with its taxonomic habits, calls "compatibilism" is a more convincing approach to free will: while the latter is not an object like a star or a tree but a mental perception, it cannot be magically "deleted," it is socially necessary, it has very tangible consequences on the lives of individuals and groups, it is not as "free" as one could infer from the very use of the adjective, and, in last analysis, it is not incompatible with chemical and physical determination. The absolutism of determination cannot content itself with exceptionalism, out of fear of conceding too much to the "folk" position and being at odds with itself. But please note that even Harris is reintroducing responsibility. Yes, he says, the perpetrators of the crime are determined, but we can still hold them accountable. So, if I understand correctly, they are irresponsibly responsible. Why not? But this does not look like a very consistent conclusion for someone who speaks of absolute, scientific determination. In the context of enunciation, the incoherence is even stronger, since the writing, publishing, and debating of such a book can only make sense if knowing that we have no free will will free ourselves. Which makes no sense.[5] No more sense, at any rate, than ignoring that we are puppets. In other words: who cares?

The only reason why we could ever care is if exceptions are granted. Harris's dilemma is just one form of any such absolute determinism. In Christian theology, if I am going to be saved, I'll be saved; if I am going to damned, I'll be damned, no matter what. But then why worship God? Why believe in this doctrine and pray? Calvin's specious response is that, through "unconditional election," God has predestined to sanctity the humans who would later be saved. So, we have to assume, the elected ones can only belong to the true church.[6] In a seemingly different perspective, Marxism participates in Determinism 1 while borrowing its theological apparatus from the second type. The laws of history and economics are implacable: they guide humanity toward freedom; meanwhile, nobody can escape them, at least before the final advent of communism. Those laws are apt to be understood "scientifically." The next questions are: Since capitalism will fail and be overcome anyway, why should we hasten the process, and, even if we do, how would completely determined creatures be of any use to the forces of history? The answer to the first question is either humanitarian (we want to spare human lives and sufferings) or utilitarian (we want to spare the means of production, or the planet). The answer to the second one is at the basis of the International. There are objective and subjective conditions to any revolution, and, through political work, one can help crystallize class consciousness. If *only* objective conditions or *only* subjective ones are met, the destruction of capitalism will not happen properly. The proletarians, who were supposed to be purely determined, can now change the society they live in. The dialectics of the objective and the subjective components will explain why rebellions fail, or when one can expect more. This has been a recurring difficulty. Marxism is certainly one of the most deterministic approaches that would not be anchored within physics, biology, or monotheism—but even there, it cannot be completely deterministic. Or it would just self-erase.[7]

I could go on. What I want to point at is that neither pure indeterminism nor determinism is tenable without making exceptions. The most radical apostles of free will end up limiting their "voluntarism"; the most absolute defenders of determinism will consider the actual and unpredictable effects of "representation"; or else the whole theoretical edifice shortly crumbles, or will be immediately deserted. This is not because truth is located somewhere in between two extreme positions. This is because absolute freedom would need to free itself from its concept to be free, and because determinism cannot be determined by itself (or it would be free). The experience of being free might well be deceptive; it will not be *replaced* by the experience of being determined, for both conceptions structurally rely on their own logical impossibilities that, nonetheless, cannot prevent their realization. In simpler terms, I am not *more* determined, *in principle,* than free.

SUE. I am not disagreeing . . . But how could anybody not say what you are saying?

LAURENT. I am glad we agree, but the majority of the discussion surrounding free will is attempting to eschew both incompleteness and impossibilities.

SUE. You mean the philosophical discussion.

LAURENT. Yes

SUE. Not my discussion.

LAURENT. Yes, your discussion as well, since the very moment you elaborate on the different kinds of determinism, or refer to authors like Sam Harris.

SUE. Hmm.

LAURENT. Now, we could say that we are done with this aspect of the debate. You may put more emphasis on the social impossibility of any consistent form of absolute determinism, I may stress the self-defeating structure of discourse a bit more, but we both agree on this: while unbounded free will is nowhere to be seen, we operate, in our very life, *with* this "representation," and trying to "eliminate" it in the name of science or God is a nonsense. To this, I am adding that Determinism 3 was already anticipating the impasse of the other two positions.

SUE. When you say that, despite any scientific understanding, we still operate with "free" will within our lives, I fully agree. I believe this is a structure of communication through language. I am announcing now that I want to discuss "Sue and her perception of language and free will." But how can I do this when I speak with individuals who perceive apes as having no language, no self, and no free will? I suggest we move from the general philosophical context to a more local examination.

LAURENT. Agreed.

SUE. Language is a system that embodies free will and a concept of self, because it embodies intentionality. Then, every act of communication among beings endowed with language presupposes some degree of free will. Even turning your back to someone is a sign of intent. You last went to the lab in August 2014, or months after I was deemed persona non grata at the facility. You told me that, during your very brief visit, all you could see of Elikya was her back. I am asking you: Is it because she was looking at something else, or because she didn't want to look at you? If Elikya is not an intentional being, then the question is irrelevant. If Elikya is an intentional being, then you need to understand what she does, if you're trying to interact with her. You have interacted with Panbanisha well enough, and she even once determined, with intent of her own free will, to take a picture of you. She was a subtle, deliberate, amazingly intentional person that, in many realms, went beyond what most human beings I saw in my life are able to accomplish.[8] But do you know if Elikya is intentional or not? Could she just happened to have been looking the other way when you were in the lab last year? If you look at cattle in the field, animals are grazing, and some of them don't notice you are there: Do they intend to ignore you? Is Elikya more like a cow, or more like Panbanisha? Do you know? Can you answer that question? If I am trying to create an ape with language, there is never a split second that I am not addressing this question.

LAURENT. This echoes the discussion we had in a previous dialogue about goal-oriented communication. A "goal-oriented" being would just respond to stimuli or drives, without manifesting free will.

SUE. Exactly.

LAURENT. The determined drive to eat an onion is being expressed by the organism pressing the key ONION.

SUE. That's exactly right, yes. Or this is Dr. Taglialatela's "fat finger hypothesis,"[9] thanks to which we are reaching another level of determination.

[We both laugh.]

LAURENT. True, according to the "fat finger hypothesis," no determined drive is putting Kanzi's finger on the lexigram: because of the physical determination of Kanzi's body, the very fat finger is not even at the right place.

SUE. In this "null hypothesis," the hand gesture for "give" followed by the appropriate lexigram could not make sense, since "we know" that Kanzi has no language . . .

LAURENT. There, we have two levels of non-free will and two levels of determination . . . Joke aside, I am now going to answer your question. First, I don't know Elikya well. I don't know her like I knew Panbanisha.

SUE. I want to stop you right there. I, Sue, am in this room, and so are you. If I suddenly turn and sit the other way for a long period of time while looking out through the window, you know me, and you are going to assume I did this on purpose. Now, say I go out on the street, bring back people you have never seen, tell them I want to do an experiment with them, and give them secret instructions. If you say hello to them, if you nod, they turn their back on you. I think you are going to presume that they did it on purpose, because they are human.

LAURENT. And because I live in a human society.

SUE. You can only preface your answer by saying "I don't know Elikya very well" *because* she is a nonhuman.

LAURENT. There are different degrees of intentionality in some cultures, and different practices of getting attention. In many places, looking at someone right in the eyes to begin an interaction is a rather dangerous practice. Even with humans of other groups, I might be more disoriented, so I might hesitate about the attribution of intent for some actions. Moreover, in August 2014, I was not allowed to go beyond the lobby area and I was in fact in a different space than Elikya was. The disorientation, then, is not *only* a factor of the human/nonhuman divide. My subjective answer to your question is: I thought Elikya was turning her back on me. I thought she did it on purpose. Which purpose exactly? Because she does not like me? Because she no longer trusts humans? Because she had to give the impression she was not siding with Sue's friends? I do not know.

With Nyota and Kanzi, the situation was different: while being in the same room as Elikya was, they came to the glass to look at me in the eyes and stayed there for several minutes. Nyota blew a kiss. Neither Kanzi nor I had access to a keyboard, but I could speak in English to him through the glass, and he would nod. I asked him orally if he remembered the last time I was at the lab in 2011, and he nodded. Later in the day, Kanzi saw that everybody but me had left the lobby area. He came right to the glass, and he made sure visually that the person in charge of the daily operations was outside. Then Kanzi waved toward me, making a conventional gesture to invite me to come closer. I didn't think that maybe Kanzi's fingers were . . .

SUE *[laughing].* Too fat . . .

LAURENT. And that, because they were too fat, Kanzi had to move them in what precisely happened to be my direction and that all this coincidentally occurred in the two minutes I was alone in the lobby with nobody surveilling us. I definitely put "free will" into this series of gestures, even though Kanzi is a bonobo, even though he could not use his keyboard to communicate with me.

SUE. If you assume that a being has free will, you would also assume it has intent. If "free will," in the "incomplete" but effective sense we described before, is operating most of the time, then most actions are intentional, whether it is a speech pattern, or thumping you with a stick, or glancing around and looking back at you. We cannot have the phenomena of language without the phenomena of free will and of intentionality. According to the way psychologists have traditionally looked at the world, there is a mass of organisms that are making responses to each other. One organism does something, there is a response. Another organism does something, there is another response. One can speak in terms of stimuli and responses, or conditioning and responses; those are two different psychological constructs to explain behavior. Most traits of animal behavior will be explained by these kinds of paradigms. What single cells do could be easily explained at this level, as a biological system with no nervous feedback. But let us consider quorum sensing, when cells send out chemical signals. When they sense there are enough of them, some will photoluminesce, for example. Or they somehow come together and form a big colony, whereas they had been previously distributed otherwise. Would you say the cells intend to do that? Would you say the cells have the goal of doing that? Would you say the cells . . .

LAURENT. Are programmed to do that?

SUE. Or that they have free will? Many biologists who describe cells would use this sort of a vocabulary. Psychologists wouldn't. Even if we get up to the next level, which I think characterizes most animals, with intent and expectancy, psychologists would hesitate to use this lexicon. In Skinnerian conditioning, there exists what is classically called *extinction*: a pigeon learns to peck a key and gets a reward; when you stop giving the reward, the pigeon still pecks the key, then it seems to be surprised and it looks at the machine as if it were asking where its reward is. A Skinnerian psychologist would explain the scene by saying that the pigeon is conditioned, and it doesn't do any good to add words like *intent* or *expectancy* on the part of the pigeon. You are simply trying to make an explanation at a behavioral level, you do not concern yourself with what could happen in the pigeon's mind. This position was perfectly acceptable until the invention of computers and until we had a way of opening the black box and writing the program. Earlier this year, I was told by Joseph that Skinner is not even studied at Cornell anymore. Nobody even mentions him. Because of computer science, we have opened the black box. And the black box is taken to be the model for a conscious nervous system that provides feedback. But as far as we know, computers have no self. The self is a psychological construct that the child forms across time through embodiment and the perception of having free will and free motion. It looks at its hands and feet, it is moving its hands and feet.[10] Apes do it, and humans do it. Through the embodied *exploration* of itself, the baby is executing the selfness of its self.

LAURENT. Both exploration and execution continue through language.

SUE. Yes. Because you are constructing yourself through your interaction with the other self, your self builds up as intentional and as interacting with others, whom you find to have selves like you do. As soon as you find selves like you, you construct the group; you also construct individuals who are not part of your group, and how you would feel in the absence of a group. If, by raising a child, we are going to get the emergence of humanness, it is appearing at once. We find this in both apes and humans, but, as I think I said, there are different trajectories for humans and apes. You may alter the trajectory, which captivity does by default, by taking the baby from the mother and putting it into the zoo nursery. If you alter it by intent, you may be trying to give the baby ape all of the experiences that it would have if it were human. To me, the onset of the Tarzan story is perfectly realistic. A human baby could be on a very different trajectory if it were raised by chimpanzees. Just, it would not turn into a Lord Greystoke as an adult. Impossible. But it would be a very good ape. Possibly more adept than the apes who had raised it. There would not be any lacking in this regard. If we had been allowed to continue as I desired with Nyota or Teco, they could have been very good. They have the potential to be very amazing ape–humans, in the same sense that a human might be a very amazing human–ape, while still retaining characteristics of its humanness—and Teco and Nyota would still retain characteristics of their apeness. This can take place because, in both humans and apes, there is already the capability for free will, for the self, for constructing symbolic layers, and going back and forth between them.

Most of humanity has such a tiny idea of what rearing and culture actually do. It is just horrendous that we don't understand what we are doing to ourselves with every generation. If we understood it, we could become "like gods" by conscious improvement. Instead of this, we are re-creating the same human problems over and over. This is most apparent to people in the modern Western world when they look at societies where there are lots of violent conflicts and ask, "Why do they have to have all these wars?" But they are not looking at the problems they have in their own society. How easy it would be, in one, two, or three generations, to create new conditions! For me, although it is not yet fully explicated in my own work, the message of a Kanzi, a Panbanisha, a Teco, or a Nyota is that we humans do have a hugely untapped potential that we are clamping down and confining very early on, in order to fit the preconception of humanness that is held by the society around us. We have no full concept of what we could do and could be, of the really good world we could create. That is the main message I draw from the work.[11]

LAURENT. As a postscript to this *message,* I propose we comment upon the 2007 article you published on the "Welfare of Apes in Captive Environments."[12] This essay deals with free will at many levels. It examines the type of environment that would respect the free will of the languaged apes, and be conducive of it. You asked the bonobos about their life in captivity, but you also cosigned the article with Kanzi, Panbanisha, and Nyota. This simple decision—of your own free will, I suppose—generated many hostile and enthusiastic responses. In familiar terms, you really hit a nerve . . .

SUE. This paper was done after we moved to Des Moines, and the move itself was extremely difficult. I was invited to give a talk at a conference on the best housing conditions for apes, with proceedings to be published by the *Journal of Applied Animal Welfare Science.* At that particular moment in time, I was also trying to integrate new staff at the Iowa facility. I had tried to build a new lab offering ideal housing conditions for the bonobos, but we had no outside area. There were very different views at the Great Ape Trust about what apes needed most. There were orangutans that had always been reared in zoos, as well as Kanzi and Panbanisha, who had always been reared in a forest. When I received the invitation to speak at the conference, I thought that since I had trouble convincing people locally of the real needs of these apes, I might face even more difficulties elsewhere. I had this idea: why not ask Kanzi and Panbanisha, why not try to generate real data on their own perception of their needs? I could communicate with them, they could understand these questions, and they could answer them. I asked questions to Kanzi, Panbanisha, I recorded the sessions, and I told the PR department to put the recording online, allowing anyone to access it and make up their own mind in connection with my talk and article. I immediately faced problems within my own lab. The people working with the orangutans were very upset; they did not want the video or the audio to be posted on the Web site. Others, who, technically, were working for me, did not want this information to be widely distributed either. In contrast, when I played the tape at the conference, everything was really well received. Then, of course, I was asked to give a written version of the paper.

I thought that the recordings would be helpful to people who did not believe I could even talk to Kanzi and Panbanisha. Then, when it was time to write everything up, it became very clear to me that I was pulling together the work of other coauthors on the paper. This was what I was doing. They had their opinions; I was trying to make sense of them. I was trying to put them in a framework that people could easily understand. Oftentimes, I worked that way with coauthors. So I incorporated their names in the signature. Since I had not presented the bonobos as coauthors at the conference, I did not know if it would pass peer review—which it had to do—or even if the journal editor would agree that I could do it. Once the editors and reviewers read the paper, they agreed that I was not making an illegitimate demand, because the whole theme of the conference was about the best possible environment for apes. Nobody could really ask, "Could the apes understand your questions, and were they really answering?" This was overwhelmingly clear. I instituted the control condition where, instead of asking the bonobos my questions, I read them segments of *Science* magazine to see if they would respond. They didn't! The accuracy of my understanding was a different issue that had nothing to do with whether Kanzi or Panbanisha deserved to be coauthors or not. They spoke, they talked, they listened, and they answered. In my view, this gave them the right of authorship. The article was accepted for publication. But even *before* it was published, I knew this was going to be a problem because there was outrage among people at the Trust. It did not last long before they succeeded in removing me. Many people felt I had crossed the boundary of what a sane scientist would do. The article was a very difficult one to write: no insane scientist would have been able to write and present it! At the conference, as I attributed all this information to Kanzi and Panbanisha, no one stood up and said, "We don't think they're really doing this," or "We're questioning if they really understood your questions," or "We don't like your framework." There were both zoo people and scientists at the symposium, some being real critics of my work, but everyone accepted what I was presenting, although they might have questions about certain aspects. The overall format was not challenged. By putting the names of Kanzi, Panbanisha, and Nyota on the paper, I created this tremendous anger in my own institution. Was it because the orangutans, who were then also living at the Trust, were unable to answer similar questions, while the PR department insisted on presenting them on equal footing with the bonobos in terms of symbolic competence? I don't know.

LAURENT. The outrage passed the confines of your laboratory. One source for that scandal was tied to the very issue of free will, and to the practical consequences for ape welfare, if we were to ask chimpanzees about their preferences. But don't you think that part of the uproar was also about a psychologist listing her "subjects" as coauthors?

SUE. Yes, if you are thinking along the line of *I am a scientist and I am interviewing these people.* This wasn't the format. The conference had a pragmatic approach: How do we know what environments are best for animals? The general thought is that we cannot interview animals. In a sense, I viewed myself as an interlocutor for Kanzi and

Panbanisha, allowing them to speak to other scientists about their best environment. I was not an interviewer. This makes a difference.

LAURENT. Assuredly, but a scientist is not supposed to be in this position. The scientist is usually an observer.

SUE. Yes, usually. Let us say that I am an anthropologist writing to my peers about this nation I studied and lived with for an extended period of time. I might not include the name of all members of the tribe as coauthors. But let us say I go back to the tribe. We are trying to get water and sanitary facilities in this village, and the people are trying to talk to the outside world about how they want to keep the village theirs while also having these modern facilities. They are having trouble talking to the outside world, but they are making progress in doing it. Now, I go to a conference about what it means for indigenous societies to be moving into the "modern world," and I am a vehicle to help them go into this direction without being decimated the way they often are when they have Western contact. I am more than a scientist, I am working with these people, hand in hand, to make changes in their world. One of them could actually come and lecture with me. I would not cite everyone in the group, but I would cite the ones I had been working with more closely. Yes, I would put their names on the paper.

LAURENT. Sixty years ago, you would never have done that. You were the anthropologist and you had informants that perhaps you would mention en passant. Perhaps you would not even state in your monograph that almost everything you could write about this tribe was coming from one or two persons, who accepted to talk to you for hours.

SUE. True.

LAURENT. The official ethics of anthropology at one point was that you would hide all this contact.

SUE. Yes.

LAURENT. Then the question changed. As Claude Lévi-Strauss did in *Tristes tropiques,* you could relate your experience in a diary or a narrative—but this would be separate enough from the scientific monograph, where you'd largely conceal the fabric of life and your own involvement in it. Then, in a newly revised version of the discipline, including "indigenous anthropology," one finds the idea that if you work closely with a "native informant," this person could come with you to a conference or be identified as a coauthor. At the very least, you would stress the role of the people you worked with. I believe that, by putting the names of the bonobos as coauthors, you were advocating, in the study of great apes, something closer to that shift within anthropology. But, by doing this, you were also breaking a professional rule in the fields of primatology or psychology.

SUE. Yes. Except that I was presenting an interview, and experimental psychology has no place for this. Interviews fall within the domain of clinical psychology. My approach would probably not have caused the same ruckus if it had been published in a clinical journal.

LAURENT. This drives us back to the question of free will, which certainly was the main reason for outrage. In writing about that topic of welfare of captive apes, given both the goals you had in mind and what you believe in, you said it was almost impossible for you not to list the bonobos as coauthors: if you had not done this, you would have denied them the free will that was the very basis of the question.

SUE. That's right.

LAURENT. This coauthorship, then, was a performative gesture. I think that many people, including those who reacted negatively to your article, fully understood the situation. What you did was obviously very powerful or disturbing: you inscribed this possibility of free will in nonhuman agents.

SUE. That's right. I would add that, yes, I followed this progression in anthropology. The current view among anthropologists is often that we are not even going to go to any of these indigenous societies anymore, because this is such a colonial viewpoint. Besides, there are many indigenous people now who have gotten a college education and said "I know more about my society than you do." This is like Panbanisha going to school and saying "I know far more about being a bonobo than you and you have written all these papers: I'm tired of it, so I'm going to tell the real truth!"

I have sought funding to eventually return this group of bonobos to the Congo and allow them to interact with other members of their species, not because I consider they need to become bonobos or indigenous again, and not because they no longer need human contact. I simply know they should not be in a cage. I could create an environment without a cage and explain to them why they could not go beyond the boundaries of their environment, how dangerous it was, how people viewed them, and why they had to stay in it. I could make it a wonderful environment so that they didn't really mind staying in it any more than anybody minds staying in Osceola, Missouri. But it is not legal to do such a thing in America. You must have the apes in a cage because, under federal law, they are considered to be a wild, endangered animal species. Because these bonobos have free will and a concept of self, because they would like to engage in a reasonable amount of self-determination, and because they wish to continue some kind of dialogue with human beings that incorporates those features, the only place that is available for them to *flourish* is the Congo, where there are other bonobos that are not caged. I feel as though I am personally on the fast track that cultural anthropologists have gone through, over several generations and decades. I am inevitably moving in the same direction.

I would not like to think of a future in which there are no "contacted bonobos" and we never bother to learn anything more about them, about who they are and what their culture is. Although I concede that "contacting" bonobos may be harmful to them, we have such an extraordinary amount to learn from further investigations on the appropriate basis of communicative equality that staying away from them would be a great mistake for humanity—and for the bonobos as well. Of course, you can't assure in the Congo the same amount of safety that you can in an 8 x 10 cage that is cleaned and disinfected daily. You can assure a certain amount of safety in a cage, as well as a vast amount of unhappiness, even destroying the desire to live in such an environment. But it is very safe! Even though no human wants to live in it, it is very safe!

LAURENT. All this reminds us that free will is not, and cannot be, a nonpolitical category.

SUE. Yes.

LAURENT. Here, a codicil: in the last few decades, a growing number of primatologists, for potentially idealistic or more pragmatic reasons, became gradually convinced that it was not ideal to house apes in labs, because of the legal issue surrounding caging and also of fear of animal-rights activists. Observing ape societies in the wild became the way to go. Seeing groups of bonobos from afar without interaction is what you describe as "noncontact." There is a quest for purity that stands opposite to what happened in ethnology, where the current focus is no longer on the museum-like concept of the Other.

SUE. We do need to study ape societies as they are "in the wild," but we should not consider their cultures to be static and nonhistorical, or "naturally" determined. Cultural changes happen across time for chimpanzee cultures, which opens the door to more transformation through "contact." Itai Roffman's observations of wild chimpanzees in Mali, for instance, validate, corroborate, and expand the findings achieved by the language studies.[13]

The thing that gives the human or the ape the sense of free will and lack of restriction is being able to be part of a body politic that is acting as a culture or a group. Inasmuch as this culture or group is making its own decisions, it has, to a certain extent, its own free will, and you, as a member of that group, have a role to play in decision making. This group has to be earning its own way in the world. Bonobos do it in the wild: they decide where to go, they get their own food, they have their own rules and regulations for how they treat each other and how they treat other groups of bonobos. Such a level of self-determination could only be granted to the groups of bonobos I raised if they were in *the* country on the planet where other bonobos exist and currently have, by default, the right to do this. This right might not last very long. It is not exclusively a matter of not being shot for food or being "conserved." It is a matter of having the human population understand that bonobos have the capability for free will, and giving them the right to exist as a sort of alternative nation.

Animals don't have rights vis-à-vis each other. Only humans are assumed to have the conscious ability to make laws, hence to promulgate duties coming with rights. When it is said that humans must develop laws for each other and extend them to animals, they are not extending those to animals in the way they are extending them to each other. If you have an organism that can develop language, that can recognize itself in a mirror, that can develop a moral system, then you must give that organism the right to develop a moral system. You must give it an environment and a place to live, as well as opportunities for it to establish its own moral code for how those beings treat each other. If this impinges upon human rights or human life, you must discuss with this being the problems that such choices entail. You must expect some moral give-and-take to occur between those two beings through a dialogue. Giving an ape some right to live in a larger cage, or to be with conspecifics, really, that's not it! The ape should have the right to exist across generations, to form its own culture, and to formulate its own moral rules. If this is not done, then it still is another species (humans) that retains the ability to parcel out rights. You have to let that other species determine itself, just the way we need to let other nations make up their own choices, and certain indigenous people within certain nations to do the same, of their own accord. This is the kind of rights that apes require!

LAURENT. And, typically, this is not the kind of rights that are being re—

SUE. Requested.

LAURENT. Requested now.

SUE. If we let go and stop the kind of work I have done, we will not only leave the bonobos behind culturally, but also, in the fog of time, they will morph into our concept of "the animal." Having realized they do not belong to such a category, I see this possibility as an immense moral wrong that the human species would forever regret. We have a moral obligation to understand, and the only ethical way to proceed is to allow for situations that preserve the apes' capacity to manifest their potentialities, should they have them. If there are other beings that have free will, a self, and are interested in self-determination, and if they are capable of moral systems—whether they are whales, elephants, dolphins, or apes—we have an obligation to understand them and find some opportunity for the consciousness of those species to continue to exist. If we don't do this, we are turning our back on another manifestation of sentient consciousness, and we are not understanding our own role. If we cast apes away, we will diminish ourselves, because we will be making the wrong judgment about them. And we will not understand ourselves if we make this mistake. We will protect a current limitation in us that we would not really want

> to be protecting, and of which we are even currently unaware. We haven't yet achieved the level of understanding that would allow us to stop. What I just said certainly presumes something about where humanity is headed. I acknowledge that.

Greek "myths" have filled my life since I discovered them at the age of seven. In ancient tragedies, human actions are superseded by another, godly, logic of Necessity. But those plays do not only show the inexorable. They repeatedly remind us that, even if the end is written, even if all countering efforts are ultimately futile, and even if death will come upon us and our dreams, we also have to determine ourselves. Moreover, the very journey leading to the inevitable is what is worth telling. Once the Greek gods and their soothsayers are set aside, other—biological, social, political, legal—commanding principles and rational prophecies could be said to replace them. I am not so interested in those Paul Celan ironically named the "siphets and probyls" of our time.[14] I believe that whatever reality determinism could have in our lives, the latter are only worth remembering if they at least tried to transgress the rules of the "given" and to create. The immediate future of the bonobos might look bleak, Sue's research with the apes might be upended once and for all. In fact, nobody knows any of this, so, now as ever, it is vital to emphasize the significance of her work and her fight. As for the intellectual and existential journey of all those who have been and still are involved in such a transformative experiment, it is worth remembering—understanding—emulating. Thus, our last dialogue was not final; it is to be continued.

Appendix

A Timeline of Ape Language Research[1]

1661

Samuel Pepys observes a living African ape and notes in his diary: "I am of the mind it might be taught to speak or make signs."

1735–62

In his *Systema naturae,* Carl Linnaeus classifies the ape that Jacob de Bondt named "orangutan" in the *Homo* genus. Linnaeus affirms it is nocturnal and "speaks by whistling."

1748

In *The Man-Machine,* the French philosopher Julien Offray de la Mettrie makes the hypothesis that it should be possible to teach "the great Ape" how to speak by applying the method Johann Konrad Amman invented for human deaf children to utter words.

1774

In *Of the Origin of Language and Progress,* James Burnett Monboddo proposes to make an experiment with "orangutans" and to teach them sign language.

1874

In an article for *The Academy,* Max Müller, a professor of comparative philology at Oxford University, compiles information on several recently discovered "wolf-children" in India. One of those feral children is quadrupedal, refuses all clothing or cooked food, and never acquired language. Müller notes that "the cases of children reared by wolves afford the only experimental test for determining whether language is an hereditary instinct or not."

1896

In *Gorillas and Chimpanzees,* the colonial adventurer and naturalist Richard Lynch Garner reports that he tried to teach Moses, a one-year-old male chimpanzee, to utter a few monosyllabic words (borrowed from different human languages, including German, French, and Nkami). The instruction lasts for only three months, as Moses dies from fever. Garner claims that, with more time, the ape "would have mastered these and other words of human speech to the satisfaction of the most exacting linguist." Garner's further attempts, in particular with a female chimpanzee named Susie during the 1900s, are largely unscientific and inconclusive.

1909

Lightner Witmer, a professor of psychology at the University of Pennsylvania and one of the founders of the field of "clinical psychology" in the United States, observes Peter, a young chimpanzee who performs a music-hall routine that inspires both Franz Kafka ("A Report to an Academy," 1917) and Buster Keaton (*The Playhouse,* 1921). Witmer subjects Peter to the cognitive tests he uses with children having mental retardation and, in one session, he teaches the chimpanzee to say *mama.* Later, William Furness III works under Witmer's guidance and teaches a female orangutan to utter two words, *cup* and *papa.*

1915

Luis Montané reports the birth of Anumá, who is considered to be the first chimpanzee born in captivity in the West. His parents live in an ape nursery, established in Havana in 1906 by a wealthy woman called "Madame" Rosalía Abreu. Her facility later inspires several scientists, including Robert Yerkes, who opens his primate research station in Florida in 1930.

1917

Wolfgang Köhler—a German psychologist, who has worked at the Anthropoid Station of the Prussian Academy of Sciences, located in Tenerife—demonstrates in *The Mentality of Apes* that chimpanzees are able to solve problems mentally sans trial and error, the method employed by all other animals tested up to that time, as claimed by one of the founding fathers of psychology, Edward Thorndike. The latter position is still held by many psychologists and continues to serve as the theoretical basis for the rejection of ape language studies.

1925

In *Almost Human,* Robert Yerkes, then a professor of psychobiology at Yale University and a pioneer of primatology, speculates that apes might be able to be taught sign language given their high degree of nonverbal intelligence. Yerkes raises two young apes side by side and finds them to be extremely different. One, named Panzee, is a common chimpanzee, and the other, named Chim, is a bonobo. At the time, bonobos had not been identified as a separate species. Yerkes, however, identifies differences between these two individuals that later prove to be characteristic of species distinctions, the most prominent being that Chim is far more vocal, with a much wider range of sounds and frequencies. Chim's "utterances" are carefully transcribed into music. This work has never been followed up, but likely holds the key to understanding the bonobo's own symbolic mode of communication.

1926

Joseph Amrito Lal Singh publishes an account of the two children, Amala and Kamala, who have allegedly been reared by wolves. Kamala is approximately eight years old and Amala is about eighteen months. Both are quadrupedal, have no language, reject cooked food, and are nocturnal. They do not want to be dressed and scratched; they bite people who try to treat them as "human." The authenticity of the story is later disputed, but it sparks interest in the scientific community and influences Winthrop Kellogg.

1930

Luella and Winthrop Kellogg (a professor of psychology working at the Yerkes Primate Center) co-rear their son Donald with a female chimpanzee named Gua. Donald and Gua form a tight bond and Donald begins to emulate chimpanzee vocalizations and tree climbing, while Gua begins to walk bipedally. Her language comprehension and tool use develop more rapidly than Donald's. Winthrop Kellogg terminates the study after nine months, although he had first intended it to last till adolescence. He feels that he already demonstrated that cultural variables—not species variables—are determining the most important human and chimpanzee traits. Furthermore, there arises a concern that the continuation of the study has the potential to affect Donald's development in a negative manner, to a far greater extent than anyone had previously considered.

1935

Nadezhda Ladygina-Kohts, a "zoopsychologist" at the Darwin Museum in Moscow who had studied a male chimpanzee infant named Joni in the 1920s, publishes a monograph comparing the development of emotion and intelligence in ape and human children.

1948

Japan is the only country where primatologists grow up seeing primates. Kinji Imanishi, the founder of Japanese primatology, begins the systematic study of the social behavior of the Japanese macaque with a visit to Koshima Island. By 1952, the monkeys have become habituated to primatologists, and cultural transmission of behavior (potato washing) is documented among them. Imanishi and his students pilot what will become the common methods of (a) long-term observation, (b) provisioning, (c) individual identification, which has come to characterize nearly all field studies of nonhuman primates, (d) cross-generational observations, (e) kinship records. These methods lead to the realization that cultural learning plays a role in nearly all primate behavior and they have now been extended to nonprimate species.

1951

Cathy and Keith Hayes (a psychologist at the Yerkes Primate Center) raise in their home a chimpanzee they name Viki from birth to seven years of age and try to get her to speak. As described in Cathy Hayes's *The Ape in Our House,* by molding her lips, they teach her to utter the words *mama, papa, cup* and *up.* However, they find that Viki's language comprehension greatly exceeds her production abilities and occurs without intentional training and/or trial-based rewards. Perhaps even more important than her receptive language abilities is the observation that Viki creates, and plays with, imaginary toys. The study is terminated when young Viki dies of encephalitis. By then, Viki has learned to sew, to wash dishes, to sort photographs, and to do many other things deemed appropriate to the human female role.

1955

William B. Lemmon begins rearing two wild-caught chimpanzees, Pan and Wendy, after reading about Robert Yerkes's work with the apes Panzee and Chim and the chimpanzee colony Yerkes created in Florida. Lemmon begins a similar colony at the University of Oklahoma. Unlike Yerkes, however, Lemmon is not a behaviorist but a clinician, and he is heavily influenced by the work of John Paul Scott with dogs.

1967

The Kyoto Primate Research Institute is established.

1968

Jane Goodall publishes a monograph titled *The Behaviour of Free-Living Chimpanzees in the Gombe Stream Reserve* and forever changes the study of ape behavior and the ethological studies of apes. Her use of the term "free-living" instead of "wild" is also prescient.

1969

Reporting in *Science Magazine,* Beatrice and Allen Gardner become the first psychologists to demonstrate that chimpanzees can acquire true productive linguistic competence with more than a few words. During the first three years of her life, the chimpanzee Washoe acquires more than seventy ASL signs and begins to combine them into simple sentences, much as a human child of the same age would do. Washoe, having been wild caught at approximately nine months of age, lags somewhat behind a normal child, possibly because of an absence of prenatal exposure and delayed postnatal exposure to language properly contexted. Washoe also begins to form simple two- and three-word sentences, raising a vigorous debate about whether she "has" syntax or not.

1970

The Gardners terminate their initial study and send Washoe to Lemmon's institute in Oklahoma. Roger Fouts, one of their graduate students, goes with her and takes a position at the University of Oklahoma.

Sue Savage-Rumbaugh, who has been admitted to Harvard and is on her way to study with B. F. Skinner, happens to stay with a friend in Norman, Oklahoma. She meets Roger Fouts, who invites her to the "Chimp Farm." After spending three full days with the apes, she decides to reject the offer from Harvard, as Skinner has overlooked chimpanzees and the role of language in their development when he formulated and refined his theories of human behavior. When Sue begins her graduate study at the University of Oklahoma, Lemmon's facility has more than thirty-five chimpanzees, as well as *Macaca nemestrina, Macaca arctoides,* gibbons, a siamang, a baboon, peacock, pigs. Five chimpanzees have been placed in human homes from birth, and their development is carefully monitored. Savage-Rumbaugh works closely with Roger Fouts, Roger Mellgren, William Lemmon, and Jack Kanak. She focuses her studies on children, apes, verbal learning, developmental psychology, and behaviorism, while earning her way by taking care of Lucy, a chimpanzee reared from birth by Dr. and Mrs. Maurice Termerlin. Sue also assists Fouts in teaching signs to Booee, Bruno, Cindy, and Thelma, and she daily observes the mother/infant groups consisting of Mimi and her son Meshak, Carrie and her son Shadrack, Cindy and daughter Abigail, and Pampy and her son Manny.

1971

David Premack reports in *Science Magazine* that a chimpanzee named Sarah has a reading and writing vocabulary of 130 words and that her understanding goes beyond the meaning of words and includes the concepts of class and sentence structure. Unlike Washoe, and later Lana, Sarah does not tell her experimenters her thoughts or feelings. Essentially, she solves match-to-sample problems by selecting the correct piece of plastic and placing it between two others placed in front of her by the experimenter. These two pieces of plastic are said to be "words," as is the piece of plastic Sarah chooses. It is unclear how "wordness" becomes assigned to and/or endowed to the piece of plastic, much less specific meanings, concepts, or grammatical structures, because dialogue between human and ape plays no part in Sarah's life.

1972

Sue Savage-Rumbaugh becomes Lucy's primary companion and begins to participate in her daily life.

1973

Duane Rumbaugh reports in *Science Magazine* that a chimpanzee, Lana, is able to acquire a written language with a phrase-structure grammar, with the assistance of a computer. Having broken the language barrier, Lana is able to make novel, syntactically and contextually appropriate sentences, and to engage in dialogue. In contrast to the Gardners' and Premack's efforts, all interactions are recorded on a PDP 8 computer. Thus, there is no "guesswork" as to what Lana did or did not do. Similarly, there is virtually no chance that subtle cueing effects are determining Lana's behavior, because she can perform quite well when no one is in the room and she is "communicating" only with a computer. Like Washoe, Lana has acquired these skills rather quickly, thus suggesting that the barriers of previous experiments (such as the Hayes') were the result of trying to get the chimpanzees to speak. Both signing and symbolic reading are a means around the differences in our vocal tracts and that of the chimpanzee. Lana does not learn by molding and shaping: she only needs to press a key.

Roger Fouts demonstrates that Washoe's acquisition of sign language is not unique, a position that—strangely enough—many critiques took at the time. In an article published in *Science Magazine,* Fouts reports that Cindy, Thelma, Booee, and Bruno (all aged between four and six) are able to acquire ten signs via the method of molding and shaping. In this method, ostensibly invented by Fouts, Washoe's hands are placed in the proper configuration for the sign while the object is held in front of her. Then, shaping techniques are employed to gradually reduce the amount of molding needed until Washoe produces the sign without assistance. Prior to the implementation of this technique Washoe's sign acquisition was slow, similar to that of Viki, because the Gardners were relying on the process of imitation. Washoe can imitate new signs but does not do so frequently.

Onan (son of Mona) is born and reared for the first three months of his life by Sue Savage-Rumbaugh in her University of Oklahoma student apartment, with her son Shane. Onan is later sold to Herbert Terrace, who renames him Nim Chimpsky.

Toshisada Nishida reports that chimpanzees in the Mahale Mountain use tools to gather ants, and the first indications of cultural differences between groups of wild chimpanzees begin to emerge.

1975

Sue Savage-Rumbaugh completes her PhD and moves for a postdoctoral fellowship to Atlanta (where the Yerkes Center has been relocated since 1965). She begins comparative behavioral studies of bonobos and chimpanzees, focusing on three bonobos who had just arrived from the Congo: Bosondjo (male of six years), Lokalema (female of sixty-plus years), and Matata (female of five years). She compares them to a group of wild-caught chimpanzees of similar age/sex composition.

1976

The New York Academy of Sciences annual meeting is dedicated to ape language and organized by Horst Steklis and Stevan Harnad. Herbert Terrace, who teaches psychology at Columbia University and had mainly worked with pigeons, announces that he is going to finally do ape language research the way it should be done with his subject Nim Chimpsky. He insinuates that the Gardners and Rumbaugh were not sufficiently knowledgeable of behavioral-shaping techniques to produce the correct results.

Duane Rumbaugh adds four young chimpanzees to the Lana

project: Sherman, Austin, Erika, and Kenton. Unfortunately, these second-generation subjects fail to follow in the footsteps of their elder group member, Lana, because of the lack of social interaction when words are acquired. Liz Rupert-Pugh and Sally Boysen join Sue Savage-Rumbaugh at Yerkes.

1978

Francine "Penny" Patterson, a graduate student at Stanford University, reports that the female gorilla Koko is also able to acquire signs by using Fouts's method of molding and shaping. Koko was born in a zoo and placed in a nursery because her mother ostensibly failed to care for her. After thirty months of training, Koko acquires one hundred signs and makes sign combinations up to eleven signs in length. Patterson purchases Koko from the zoo and forms a nonprofit corporation to fund the work.

Lyn Miles begins to teach signs to Chantek, a nine-month-old orangutan, by employing the Fouts method of molding and shaping. Like the chimpanzees and the gorilla before him, Chantek acquires signs and begins to combine them. Chantek develops a vocabulary of 150 signs that he combines spontaneously, inventing many of his own signs. This work is unfortunately terminated early, when, at the age of seven, Chantek escapes his compound on the University of Tennessee campus and begins to disrobe a coed. The university administrators are not nearly as understanding of this behavior as they are when freshman males engage in similar actions, and Chantek is summarily banished to the Yerkes Primate Center. He later finds another home at the Atlanta Zoo (when the Yerkes Center removes all orangutans from its research facilities). Chantek, like Viki, Gua, and Koko, displays comprehension of human spoken language. (Washoe and Lana did not, because humans were not allowed to speak around them.)

The chimpanzees reared in Atlanta by Sue Savage-Rumbaugh and her collaborators are now not only communicating all day with their caretakers but also beginning to use lexigrams to communicate with each other and to share novel information deliberately. An article documenting this behavior is published in *Science Magazine* and draws the ire of the archbehaviorist B. F. Skinner, much as the combined work of the Gardners, Rumbaugh,

Premack, Miles, and Patterson draws the ire of his protégé Herbert Terrace. Both set about disproving the findings of those in the field of ape language. Although they clothe their work in the mantle of "searching for the true explanations" of this behavior, as their work progresses, their agenda becomes increasingly clear.

1978

After a visit to the Yerkes Language Laboratory, Sony executives decide to fund a program of ape language in Japan, to be led by Kiyoko Murofushi. Murofushi selects Tetsuro Matsuzawa as the graduate student to carry the work forward in Japan. Matsuzawa later travels to the laboratory of David Premack and spends a year there before returning to set up a behavioral system that is designed to combine what he believes to be the "best aspects" of Premack's and Rumbaugh's approaches. Namely, he presents a variety of match-to-sample tasks on a computer screen that employ symbols similar to the lexigrams used by Lana, rather than the plastic chips employed by Premack. While Sony funds the Kyoto language research program in Japan, it also makes an important contribution to Rumbaugh's laboratory in the United States.

1979

Terrace publishes an article in *Science Magazine* detailing his findings with Nim. He claims that while Washoe may have learned some words, she has no idea of what a sentence is and neither does Nim. Their sentences are said to be formed by one-word expressions of desire coupled with "wild cards" words (such as *you, me, give, there,* etc.) that could basically be fit into any utterance. Terrace's attack forces the Gardners to reveal that they did not carefully record word order when they reported on Washoe's combinations. This article sends shock waves through the field of psychology and, from that point on, *Science Magazine* refuses to consider any other publications in the field of ape language, stating that those debates must now take place within psychology journals. *Science* does not ban other animal behavior studies from its pages, just ape language. In so doing, it serves a role not unlike that of the Linguistic Society of Paris when it decided in 1866 to ban all studies of language origins.

Kiyoko Murofushi invites Duane Rumbaugh and Sue Savage-Rumbaugh to lecture at the Kyoto Institute for Primate Studies.

1980

Drawing on a combination of ape language studies (especially the work involving Koko), of *King Kong,* of *The Planet of the Apes,* of adventure novels, and of Kafka's "Report," best-seller writer Michael Crichton publishes *Congo,* a science-fiction narrative featuring gorillas being taught sign language. Over the next few decades, the scientific research on the use of human language by great apes becomes a recurring topic in what forms a fictional subgenre illustrated by novels such as Daniel Quinn's *Ishmael* (1992), Sara Gruen's *Ape House* (2010), Tristan Garcia's *Mémoires de la jungle* (2010), Benjamin Hale's *The Evolution of Bruno Littlemore* (2011), and movie features such as *Project X* (1987), *Congo* (1995), *Rise of the Planet of the Apes* (2011).

The annual meeting of the New York Academy of Sciences becomes a bully pulpit for Thomas Sebeok and Herbert Terrace, who denounce the field of ape language as fraudulent. Terrace has entered it and failed; therefore, the success of others has to be a mistake. Duane Rumbaugh is referenced as "Dumbaugh the flying chimpanzee." The magician James Randi is brought forward to illustrate how easy it is to fool people on purpose. Having received wind of this mal intent, the Gardners stay away from the meeting, while the Rumbaughs decide to go and confront the critics directly. The critiques offered at the meeting are not even directed at the actual results of ape language reports but rather at the "silliness of the idea" that apes could talk.

The Gardners begin a second round of ape language studies with four infant chimpanzees, Moja, Dar, Pili, and Tatu. Each chimpanzee has its own house and its own set of graduate students as caretakers. During daytime, the chimpanzees and caretakers come together for language training sessions and play sessions. This is very different from the "sole subject" life experienced by Washoe. The Gardners carefully record word order as sign combinations begin to appear. Concerns about cueing have, however, become intense and signing, as a method, does not lend itself to controls for cueing in situations of natural dialogue.

It also does not lend itself to clear sign interpretations. Because of their inability to address these latter issues satisfactorily with the site visitors, the Gardners eventually lose their funding and send their chimpanzees to join Roger Fouts, Washoe, and Loulis at Washington State University.

Kanzi is born to Lorel (a bonobo owned by the San Diego Zoo) at the Yerkes Field Station. He is stolen by Matata before he is thirty minutes old and nursed by her. She is already nursing her own son Akili, who is nine months old. The name Kanzi is selected by the Yerkes Center director Fred King.

1983

In a series of experiments conducted by Sue Savage-Rumbaugh, the chimpanzees Sherman and Austin are shown to be able to provide information to each other. They can, and do, make statements about their intended actions—even though these statements provide them no explicit reward. It is further demonstrated that, before the capacity to make statements about intended actions can arise, the organism must have the capacity to ask for things, to respond to the requests of others, and to label. Lana also demonstrated these capacities, plus the capacity to form sentences that were not composed of "wild cards."

1985

Sue Savage-Rumbaugh's work with Kanzi's immersion in a symbolic world and use of computerized lexigrams demonstrates that apes have the ability to acquire language spontaneously, without trial-by-trial training, shaping, or food reward. All that is required is a cultural environment, bonding, and attachments to members of the culture, as well as complete inclusion in all cultural activities as a full-fledged member of a given society. Along with language emerges spontaneous tool use, comprehension of spoken English, artistic and musical ability, capacity to make and maintain a fire, ability to swim, or, in brief, all the things that define "humanness." The issue of "what apes are able to do" no longer deals with training. What apes are able to do is, just as is the case for humans, a function of rearing and environment.

In Japan, Tetsuro Matsuzawa, Murofushi's protégé, publishes his first work that ostensibly partakes in the field of ape language with an emulation of Lana's counting and color- and object-naming studies.

1986

In her book *Ape Language,* Sue Savage-Rumbaugh offers an explication of how a being with the potential for symbolization makes the transition from behaviors that are explainable by traditional behaviorist parameters to ones that are not.

1989

Duane Rumbaugh and Sue Savage-Rumbaugh demonstrate that chimpanzees are able to summate items grouped separately and to count. Sherman becomes able to count to twenty-one and to add new numbers beyond twelve without any special instructions, training, or effort. Thus, he acquires not only the capacity to count but also the principle of counting itself, leading to the ability to count a potentially infinite array of objects. This capacity is similar to the capacity for infinite interlocking recursion that Noam Chomsky posits as the key human process underlying all linguistic competence.

1990

Milestone in the co-rearing experiment led by Sue Savage-Rumbaugh with a female bonobo (Panbanisha) and a female chimpanzee (Panzee). For the first five years of their lives, both individuals experience the same world, side by side, with the same caretakers (Sue Savage-Rumbaugh and Liz Rupert-Pugh) throughout each day. Only their prenatal experiences and their species differ. Both follow Kanzi's trajectory. Panzee is somewhat delayed and her English comprehension more concrete and circumscribed than Panbanisha's, but her tool use and maze-learning abilities exceed Panbanisha's. Overall, just as happened in the work the Kelloggs had conducted several decades earlier, the co-rearing study creates intense bonds, affects both individuals by pulling them toward a common pole, and provides incontrovertible evidence for the power of culture as the commanding player in determining the emergence of all traits previously thought to be uniquely human.

1992

Stone-tools studies begin in Sue Savage-Rumbaugh's laboratory with anthropologists Nick Toth and Kathy Schick. Kanzi begins to knap after watching Nick only a few times. Apart from Panbanisha and Kanzi, other apes have not begun to knap.

The first computerized touch-screen technology is developed in Atlanta, allowing an even easier use of the lexigrams by the bonobos and the people working with them.

1993

Michael Tomasello and Sue Savage-Rumbaugh collaborate on a study that demonstrates the existence of the capacity to imitate in Panbanisha and Panzee. Their skills are very similar to those of human children, and very dissimilar to those of "uneducated" chimpanzees. Tomasello will later reverse his view of what chimpanzees are able to do vis-à-vis imitation while working with chimpanzees reared as pets in German homes.

In *Language Comprehension in Ape and Child,* Sue Savage-Rumbaugh demonstrates that Kanzi's understanding of spoken English is on a par with that of a human child.

NHK, Japan's national public broadcasting company, produces a movie titled *Kanzi: An Ape of Genius* documenting aspects of the life of the bonobos in Atlanta. Two other documentary films are later produced by NHK, all contributing to the large international media exposure of Sue Savage-Rumbaugh's line of research. Sue Savage-Rumbaugh works with Roger Lewin on the book *Kanzi: The Ape at the Brink of the Human Mind,* geared at a general audience and published in 1994.

1996

A *Science Magazine* article appears announcing the formation of the Max Planck Institute for Evolutionary Anthropology, with a psychology department to be led by Michael Tomasello. This is the first article to address apes and language in *Science* since the publication of Terrace's renunciation of ape language. Kanzi is noted in the article as the immediate stimulus for these facilities.

The biomedical establishment at the Yerkes Primate Center maintains privately that it finds the Kanzi work disturbing and it calls for potentially harmful MRI testing of Kanzi and the other bonobos.

1998

Sue Savage-Rumbaugh publishes *Apes, Language, and the Human Mind,* coauthored with linguists Stuart Shanker and Talbot Taylor.

1999

Birth of Nyota, son of Panbanisha and Nathan. Family rearing begins, mainly involving Sue Savage-Rumbaugh and William Fields.

2005

In *Kanzi's Primal Language,* Savage-Rumbaugh and her coauthors Pär Segerdahl and William Fields stress the importance of cultural initiation in the acquisition of language. The volume develops the importance of the "*Pan/Homo* culture."

Sue Savage-Rumbaugh's bonobo colony is relocated to Des Moines, Iowa, in a private research facility funded by Ted Townsend, a local entrepreneur and philanthropist. Sue, along with some of her regular collaborators, is now employed at the Iowa Primate Learning Sanctuary, doing business as the "Great Ape Trust." The move from academia to the private lab rapidly proves to be difficult, as many different interests and players are now involved.

2008

A major flooding of the Iowa facility is a traumatic event, with toxic water infiltrating the laboratory and the apes being trapped in the lab for several days. Liz Rubert-Pugh and Sue Savage-Rumbaugh stay with the apes.

2010

Birth of Teco at the Great Ape Trust. Sue Savage-Rumbaugh, who has been marginalized in her organization, is now called back and asked to oversee Teco's rearing, without being granted all the necessary authority. The work with Teco offers the possibility of continuing the multigenerational study. While Kanzi may not have made it all the way to full humanness in a single generation, he raises the clear possibility of the capacity of apes to do so if the forces of human enculturation are allowed to operate across generations.

Laurent Dubreuil, who, thanks to a Mellon fellowship, is pursuing research and training in cognitive science, is getting in touch with Savage-Rumbaugh, Fields, and other employees of the Great Ape Trust. Dubreuil is staying in Des Moines for a week in September and is interacting with the bonobos for the first time.

2011

Sue Savage-Rumbaugh is included in the list of the one hundred most influential people by *Time* magazine. Ted Townsend announces that he will stop funding the Great Ape Trust.

2012

In the turmoil caused by the end of Townsend's financial support and the downsizing of operations, former employees and associates of the Great Ape Trust level a series of accusations against Savage-Rumbaugh. As those allegations are being reviewed by a board committee, Sue is put on administrative leave. In November, Panbanisha passes away. The accusations having been dismissed by the review committee, Sue is reinstated in a facility that has been direly marked by the recent series of events.

2013

Through a settlement agreement executed in February, all bonobos (except Maisha) are now co-owned by the Iowa Primate Learning Sanctuary (IPLS) and Bonobo Hope Initiative (BHI), an organization previously created by Sue Savage-Rumbaugh.

IPLS is in charge of the daily operations, while Bonobo Hope oversees the research trajectory with Sue. This division of labor entails that all scientists (but Savage-Rumbaugh herself) resign from the Sanctuary board to join Bonobo Hope.

During the fall, in circumstances that are legally disputed, two former students of Sue's (Jared Taglialatela and William Hopkins) take control of the laboratory. Whereas in May the board of IPLS has resolved to guarantee Savage-Rumbaugh unfettered access to the bonobos for life, as of November she is banned indefinitely from the facility and can no longer see the bonobos or even exchange with them from a remote location. The "Ape Cognition and Communication Initiative" (ACCI) is the name of the new organization that succeeds the Iowa Primate Learning Sanctuary.

2014

ACCI is now the acronym for "Ape Cognition and *Conservation* Initiative," the word *communication* being dropped. The ACCI directors claim that Sue and Bonobo Hope have relinquished their authority over the scientific project. BHI scientists are only allowed in the lab for very brief and infrequent visits, despite co-ownership. Their demands to access the scientific protocols are denied in the name of confidentiality. In June, Sue is permitted to see the bonobos under close supervision for what is to date the last time. By diverse firsthand accounts, the bonobos are no longer constantly *immersed* in the symbolic world they used to live in, and linguistic interactions are now limited; physical contact is forbidden.

Sue Savage-Rumbaugh and Laurent Dubreuil start preparing their *Dialogues on the Human Ape.*

2015

Liz Rubert-Pugh stops working at the Ape Cognition and Conservation Initiative, as does her daughter Heather, who has helped rear Teco since 2012. After more than a year of unsuccessful attempts on the part of Bonobo Hope to reach a workable agreement with the other side, the only way to resolve the matter appears to be a federal trial held in May in Des Moines. Bonobo Hope and Sue are asking for the restoration of the research trajectory involving the apes. In November, the judge declares that he lacks jurisdiction and cannot adjudicate the claims of Bonobo Hope against ACCI.

For the first time in American history, a court allows legal hearings in the case of the two captive chimpanzees, Hercules and Leo, both defended by Steven Wise, who argues in favor of granting judicial personhood to great apes. A significant part of Wise's argument is based on language research work, in particular Savage-Rumbaugh's. In her ruling, the judge expresses some support for a writ of habeas corpus but states that she is bound by the precedent and therefore denies the demand. Other countries, such as Argentina, now have some forms of legal rights for apes.

Through comparisons of free or captive chimpanzees and bonobos (including Kanzi and his family), Itai Roffman shows the similarities in the use of stone stools or spears among contemporary apes and our human ancestors.

2017

Sue Savage-Rumbaugh gives hundreds of hours of video recordings documenting the life of the bonobos in Atlanta to the Division of Rare Books and Manuscripts of Cornell University. In previous years, the archive has been indexed and described by linguists William Greaves and James Benson with their student collaborators.

2018

Sue Savage-Rumbaugh is at work to create a tool for interspecies dialogue through computers that she calls the ApeNet. She and BHI are moving to state court to try once again to restore the research trajectory with the bonobos.

Kanzi, Nyota, Teco, Elikya, and Maisha are still living at the ACCI facility in Des Moines. Since Taglialatela and Hopkins took control of the facility in November 2013, they have failed to publish one article based on new research conducted with the languaged bonobos.

Laurent Dubreuil and Cathy Caruth begin a video archive compiling testimonies of individuals who crossed the "species barrier" in establishing a rapport beyond the human/nonhuman ape divide. At Cornell University, Dubreuil and Morten Christiansen are developing a multifold research project on the origin and future of culture and language, at the interface of the sciences and the humanities.

Notes

Introduction

1. Laurent Dubreuil, electronic message to Sue Savage-Rumbaugh, dated November 14, 2011.

2. I left the board of BHI in May 2016.

3. Throughout the text, words in small capital letters refer to the lexigrams used by the apes in the experiments of Sue and Duane Rumbaugh.

4. This quote is from a preliminary version of the Introduction prepared by Sue Savage-Rumbaugh.

5. Ibid.

6. Ibid.

On Animals and Apes

1. The final "s" here is a morpheme that is included in the lexigrams.

2. This description is adapted from the opening paragraphs of Laurent Dubreuil, "La grande scène des primates," *Labyrinthe* 38 (2012): 81–92.

3. See the "timeline" in the Appendix for additional, and factual, details.

4. In this respect, it seems that we still live in the times of Plato's *Hippias Major*. (LD)

5. See in particular Georges Bataille, *Prehistoric Painting: Lascaux, or the Birth of Art* (Geneva: Skira, 1955).

6. The lawyer Steven Wise is the leading force behind the ongoing battle to grant apes habeas corpus and legal personhood in the United States. See his *Rattling the Cage: Toward Legal Rights for Animals* (Cambridge: Perseus Books, 2000).

7. Michel Pastoureau, "Symbolique médiévale et moderne," *Annuaire de l'École pratique des hautes études (EPHE): Sections des sciences historiques et philologiques* 143 (2012): 205.

8. On this still relatively understudied question, see Karl von Amira, *Thierstrafen und Thierprocesse* (Innsbruck: Verlag der Wagnerschen Universitätsbuchhandlung, 1891), and Michel Pastoureau, *Symboles du Moyen Âge: Animaux, végétaux, couleurs, objets* (Paris: Le léopard d'or, 2012), 27–51.

9. The complete quote is, "Moving beasts have no understanding of what is right and what is wrong" ("bestes mues n'ont pas entendement qu'est bien ne qu'est maus"), in Philippe de Beaumanoir, *Coutumes de Beauvaisis* (Paris: Picard, 1900), 2:481*; trans. LD.*

10. See Raphael Sealey, "Athenaion Politeia 57.4: Trial of Animals and Inanimate Objects for Homicide," *Classical Quarterly* 56:2 (2006): 475–85.

11. See David Lewis-Williams and Thomas Dowson, "The Signs of All Times: Entoptic Phenomena in Upper Paleolithic Art," *Current Anthropology* 29:2 (1988): 201–45, as well as Jean Clottes and David Lewis-Williams, *Les chamanes de la préhistoire: Transe et magie dans les grottes ornées* (Paris: Maison des roches, 2001).

12. A recent study of the nutritional aspects of cannibalism tends to confirm that the practice is essentially symbolic; see James Cole, "Assessing the Calorific Significance of Episodes of Human Cannibalism in the Paleolithic," *Scientific Reports* 7 (2017).

13. Jane Goodall, "My Life among Wild Chimpanzees," *National Geographic* (August 1963): 272–308.

14. Sue Savage-Rumbaugh and her collaborators created what they routinely called "the Mythology" thanks to typical characters that were both impersonated, through the use of costumes and masks, and spoken about over discussions. Partial descriptions of the Mythology could be found, for instance, in Sue Savage-Rumbaugh et al., "Culture Prefigures Cognition in *Pan/Homo* Bonobos," *Theoria* 54 (2005): 311–28. (LD)

15. Self-medication in apes living "in the wild" has now been copiously documented. For a first overview, see Michael Huffman, "Current Evidence for Self-Medication in Primates: A Multidisciplinary Perspective," *Yearbook of Physical Anthropology* 40 (1997): 171–200.

16. See also what Jacques Derrida hints at in his critique of Jacques Lacan, in *L'animal que donc je suis* (Paris: Galilée, 2006), 183, as well as Giorgio Agamben's fitting argument on anthropogenesis in *L'aperto*

(Turin: Bollati Boringhieri, 2002), §§7 and 17, especially. Of course, "fitting" does not imply that I agree with Agamben's Heideggerian metaphysics. (LD)

17. "Our plaintiffs will be animals for whom there is clear scientific evidence of such complex cognitive abilities as self-awareness and autonomy. Currently that evidence exists for elephants, dolphins and whales, and all four species of great apes. So, for the foreseeable future, our plaintiffs are likely to come from these three groups." Quoted from the 2016 FAQ section of the Web site for the Nonhuman Rights Projects, headed by Steven Wise: https://web.archive.org/web/20161022065224/http://www.nonhumanrightsproject.org/qa-about-the-nonhuman-rights-project/.

18. See Cora Diamond, "Eating Meat and Eating People," *Philosophy* 53 (1978): 470.

19. W. H. Auden, "MALIN was thinking: / No chimpanzee / Thinks it thinks," in *Collected Poems* (New York: Modern Library, 2007), 452. (LD)

20. For a proposition in favor of including chimpanzees in the *Homo* genus, see, for instance, Derek Wildman et al., "Implications of Natural Selection in Shaping 99.4% Nonsynonymous DNA Identity between Humans and Chimpanzees: Enlarging Genus *Homo*," *Proceedings of the National Academy of Sciences* 100:12 (2003): 7181–88.

21. See Carl Linnaeus, *Systema naturae* (Stockholm: Salvius, 1758), 1:24; Georges-Louis Leclerc de Buffon, *Histoire naturelle, générale et particulière, avec la description du cabinet du roi* (Paris: Imprimerie nationale, 1749), 2:439.

On Dialogue and Consciousness

1. Plato, *Sophist,* 263e.

2. Julian Jaynes, *The Origin of Consciousness in the Break-Down of the Bicameral Mind* (Boston: Houghton Mifflin, 1990), book 1, chapter 3.

3. See the slightly ironical assertion of David Chalmers, *The Conscious Mind: In Search of a Fundamental Theory* (Oxford: Oxford University Press, 1997), 293.

4. See, for instance, Antoine Lutz, John Dunne, and Richard Davidson, "Meditation and the Neuroscience of Consciousness: An Introduction," in Philip David Zelazo, Morris Moscovitch, and Evan Thompson, eds., *The Cambridge Handbook of Consciousness* (Cambridge, UK: Cambridge University Press, 2007), 499–552.

5. Claude Shannon and John McCarthy in their "Preface" to their coedited *Automata Studies* (Princeton, N.J.: Princeton University Press, 1956), vi.

6. Andrew Lock, *The Guided Reinvention of Language* (New York: Academic Press, 1980).

7. See Sue Savage-Rumbaugh and William Fields, "The Evolution and the Rise of Human Language: Carry the Baby," in Christopher Henshilwood and Francesco d'Errico, eds., *Homo Symbolicus: The Dawn of Language, Imagination and Spirituality* (Amsterdam: Benjamins, 2011), 13–48.

8. See Sue Savage-Rumbaugh, Pär Segerdahl, and William Fields, "Individual Differences in Language Competencies in Apes Resulting from Unique Rearing Conditions Imposed by Different First Epistemologies," in Laura Namy, ed., *Symbol Use and Symbolic Representation: Developmental and Comparative Perspectives* (Mahwah, N.J.: Lawrence Erlbaum Associates, 2005), 199–219.

9. In the current conditions at the ACCI facility, such a possibility is being methodically ruled out in advance. (SSR)

10. Laurent Dubreuil, *The Intellective Space: Thinking beyond Cognition* (Minneapolis: University of Minnesota Press, 2015), §§91–107.

On the Flavors of Consciousness

1. In Skinner's *Walden II,* control would happen through beneficent means and for the greater good of all. As a graduate student in the early 1970s, I became greatly enamored with this idea and with Skinner himself. I believed society needed such a beneficent director, and that, nonetheless, there were inherent problems in any kind of control. Humans will *act* to stop *any* control they feel is exerted over them, including positive control for their own benefit. The mere *idea* of being controlled by others is anathema to the human spirit. Therefore, the only sustainable political system is democracy. We should do far more to protect and respect the democratic process than we are currently doing in the United States. Many people who intuitively feel this became attracted to Donald Trump—even though they strongly disagreed with his positions on specific topics. These people wanted the United States to be more democratic, to be great, to be less the backer of nondemocratic countries, etc., and this is what Trump seemed to offer to them. (SSR)

2. This is the thesis of Michel Henry in *The Genealogy of Psycho-*

analysis, trans. Douglas Brick (Palo Alto, Calif.: Stanford University Press, 1993). (LD)

3. "The ego . . . is not even master in its own house, but must content itself with scanty information of what is going on unconsciously in its mind [the original reads *Seelenleben*]." Sigmund Freud, *Standard Edition of the Complete Psychological Works* (London: Hogarth Press and the Institute of Psycho-Analysis, 1953–74), 16:284.

4. See the role of the Pink Panther in Gilles Deleuze and Félix Guattari, *A Thousand Plateaus: Capitalism and Schizophrenia*, trans. Brian Massumi (Minneapolis: University of Minnesota Press, 1987), 11. (LD)

5. A 2017 experiment requires us to update some parts of the argument we are making here. See the postscript to this dialogue for further elaboration on self-recognition in monkeys.

6. See Colwyn Trevarthen, "Infant Intersubjectivity: Research, Theory, and Clinical Applications," *Journal of Child Psychology and Psychiatry* 42:1 (2001): 4–48.

7. Peter Singer, *Animal Liberation: A New Ethics for Our Treatment of Animals* (New York: Avon, 1977), 20–22.

8. Ibid., 40–44.

9. See Cary Wolfe, *Before the Law: Humans and Other Animals in a Biopolitical Frame* (Chicago: University of Chicago Press, 2013).

10. Reference is here made to the May 2015 conference organized by the Biotechnology Law Commission of the Union Internationale des Avocats (UIA) in Minneapolis.

11. See Monica Gagliano, "Green Symphonies: A Call for Studies on Acoustic Communication in Plants," *Behavioral Ecology* 24:2 (2012): 789–96.

12. Spinoza is even more assertive in the letter LVIII to Schuller; see his *Opera* (Heidelberg: Winter, 1925), 4:266 (an electronic edition is available online at http://pm.nlx.com).

13. See, for instance, Dimitris Xygalatas, *The Burning Saints: Cognition and Culture in the Fire-Walking Rituals of the Anastenaria* (New York: Routledge, 2014).

14. See Roger Penrose, *The Emperor's New Mind* (Oxford: Oxford University Press, 1989); Stuart Hameroff and Roger Penrose, "Conscious Events as Orchestrated Space-Time Selections," *Journal of Consciousness Studies* 3:1 (1996): 36–53.

15. Liangtang Chang et al., "Spontaneous Expression of Mirror Self-Recognition in Monkeys after Learning Visual-Proprioceptive

Association for Mirror Images," *Proceedings of the National Academy of Sciences* 114:12 (March 21, 2017): 3258–63.

On Language and Apes

1. See the first canto of Nicolas Boileau's *Art poétique.*

2. "There are only two things to learn in all the languages, namely, the meaning of the words, and the grammar," Descartes says in a letter to Mersenne dated November 20, 1629, in *Œuvres complètes* (Paris: Cerf, 1897), 1:80.—*trans. LD.*

3. Ibid., 1:81.—*trans. LD.*

4. See *ibid.*, 1:80: "Establishing an order between all thoughts that may enter the human spirit, in the same way an order is naturally established between numbers."—*trans. LD.*

5. In the original, "faisant sa grammaire vniverselle pour toutes sortes de nations" means "making his grammar universal for all sorts of nations," where "he" is the (unknown) author of a project for a universal language whose merits Descartes is assessing in his letter to Mersenne (ibid., 1:79). "Universal" is an attribute to the word *grammar* in the case of an artificial language: there is no "Universal Grammar" in Descartes. *(LD)*

6. See Steven Pinker, *The Language Instinct: How the Mind Creates Language* (New York: Harper Collins, 2007), 350–51, P.S. 20.

7. Herodotus, *Histories,* 2:151–57.

8. I am using a point made in my *Intellective Space: Thinking beyond Cognition* (Minneapolis: University of Minnesota Press, 2015), §30. (LD)

9. The accusation in the name of "real science" is this: "Sue stopped doing real science by focusing entirely on rearing, she took away baby bonobos from their mothers and turned them into pets—instead of doing the kind of real scientific research she previously conducted." On this erroneous basis, that people with no or little graduate training accept as the truth, I am being excluded from the lab. Within the field of primatology itself, it is now considered absolutely reprehensible to allow a young chimpanzee to have any contact with or be cared for in any way by a human person. The only acceptable work is now to focus upon the "natural behavior" of chimpanzees—and by "natural," one needs to understand "the behavior chimpanzees engage in when they are not exposed to human beings." Anything else will be discarded and defunded, thereby losing its status as "real science." This view assumes

that apes come into the world with innate ape behavior, which may be studied. The fact that chimpanzees are as adaptable and as flexible as we are should be left unstudied. The last dialogue in this book also addresses some of those issues. (SSR)

10. William James, *The Principles of Psychology* (New York: Holt, 1890), 462.

11. See also Ernst von Glasersfeld, *Partial Memories: Sketches from an Improbable Life* (Exeter, UK: Imprint Academic, 2009), 223–28.

12. See, among other publications, Tom Mitchell et al., "Predicting Human Brain Activity Associated with the Meanings of Nouns," *Science* 320 (2008): 1191–95.

13. For more information, see *The World Atlas of Language Structures*: http://wals.info/chapter/37.

14. Luc Steels, *The Talking Heads Experiment: Origins of Words and Meanings* (Berlin: Language Science Press, 2015).

15. For all quotes: Aristotle, *Politics* 1, 1:1253a; *trans. LD.*

16. René Descartes, in a letter dated November 23, 1646, to the Marquis of Newcastle; see René Descartes, *Œuvres complètes*, 4:575; *trans. LD.*

17. See, for instance, Michael Tomasello, *Origins of Human Communication* (Cambridge: MIT Press, 2008), 114.

18. See Heidi Lyn, Patricia Greenfield, et al., "Nonhuman Primates Do Declare! A Comparison of Declarative Symbol and Gesture Use in Two Children, Two Bonobos, and A Chimpanzee," *Language and Communication* 31:1 (2011): 63–74.

19. It should be noted that Michael Tomasello recently left the Max Planck Institute and accepted a position at Duke University. After only a few years within a different context, he coauthored an article disproving a position he had tenaciously held for a very long time, namely, that there was no possibility for a "theory of mind" in great apes. See Christopher Krupenye et al., "Great Apes Anticipate That Other Individuals Will Act According to False Beliefs," *Science* 354:6308 (2016): 110–14. (LD)

20. While the specifics and implications of this measurement are widely discussed in psychology, a relatively consensual claim posits that the "mean length of utterance" among "normal" human children speaking English is a bit more than four words at the age of four. See, for instance, Mabel Rice et al., "Mean Length of Utterance Levels in 6-Month Intervals for Children 3 to 9 Years with and without Language Impairments," *Journal of Speech Language and Hearing Research* 53:2 (2010): 333–49. (LD)

On Free Will

1. "The laws of physics dictate that objects denser than water are found on the bottom of a lake, not its surface. Laws of natural selection and physics dictate that objects that move swiftly through fluids have streamlined shapes. . . . Laws of anatomy, physics, and human intentions[!] force chairs to have shapes and material that make them stable supports" (quote from "national best-seller" by Steven Pinker, *How the Mind Works* [New York: W. W. Norton, 2009], 308). (LD)

2. Léon Vandermeersch, *Les deux raisons de la pensée chinoise: Divination et idéographie* (Paris: Gallimard, 2013).

3. In a 2007 interview for the Science Studio, to the very misguided question—asked by a journalist—about what would happen "if I said in sign language to Kanzi, let us reason together," Dennett responded: "I think that's like talking to God." (The transcript is available at http://thesciencenetwork.org/media/videos/29/Transcript.pdf.) God is rather unfamiliar to me, but I am acquainted enough with Kanzi to know he could enter a rational conversation with me (and even, possibly, with Dennett). Now, I have to admit that "reasoning together" is a very odd expression, so I am not sure it would make a lot of sense to Kanzi. If the phrase is synonymous with "doing philosophy," then I am afraid that most humans would be as silent as "God." Finally, and obviously, "sign language" may not be the best way to exchange with someone well versed in the comprehension of spoken English and the use of lexigrams. (LD)

4. See the essay by Daniel Dennett, "Reflections on *Free Will*," available at https://www.samharris.org/blog/item/reflections-on-free-will.

5. "We do not change ourselves, precisely . . . but we continually influence, and are influenced by, the world around us and the world within us" (Sam Harris, *Free Will* [New York: Free Press, 2012], 63). So we do not change ourselves *precisely,* but we continually influence—i.e., imprecisely *change*—the inner and outside worlds that influence—i.e., imprecisely *change*—us? What?! Does this whole argument boil down to the differences between "precise change" and "imprecise change"? But such distinctions fall short of absolute determinism and in fact posit a level of self-determination, albeit relative. (LD)

6. "If he [God] has determined to save us, he will lead us to salvation in due time: if he has determined to damn us, we would vainly torment ourselves in seeking salvation" is, according to Calvin, the reasoning of blaspheming "pigs." Calvin's paralogical "answer" to the conundrum is that "the goal of our election is to lead a saintly life," which

apparently forbids in advance blasphemy or atheism. Quoted from Jean Calvin, *Institution de la religion chrétienne,* 24:12; *trans. LD.* (LD)

7. A very pointed critique of determinism in Marxism can be found in Cornelius Castoriadis, *Political and Social Writings* (Minneapolis: University of Minnesota Press, 1988–93). (LD)

8. Scientists like Jared Taglialatela or William Hopkins, who are now in charge of the bonobos and who, as graduate students, used to work with them, could similarly experience the free will of beings like Panbanisha. But they chose to omit their own awareness from the reports they make to other researchers, having already determined a priori that bonobos cannot do such things, or that—if they were indeed doing them at one time—this should all now come to an end. (SSR)

9. On June 3, 2014, several members of the BHI board were allowed to tour the facility for a short amount of time. According to their reports, Kanzi, as he was with Sally Coxe (the president of the Bonobo Conservation Initiative), whom he knows well, and William Zifchak (the lawyer of both Sue and Bonobo Hope) pressed BALL and SUE on his keyboard. According to Zifchak, on May 31, 2015, the exact same scene happened again, with Kanzi almost immediately asking for BALL and SUE in the presence of Sally and Bill. This second time, Jared Taglialatela was in the same space and reportedly explained that Kanzi, because of his "fat fingers," had pressed SUE by mistake, and that he only wanted to hit BALL twice. We could note that, until that point, Kanzi had used a keyboard for his entire life as well as a touch screen for decades, and that each symbol is significantly larger than the tip of even the very fat finger of a "pygmy chimpanzee." In an electronic message to Laurent dated April 26, 2017, Bill Zifchak writes: "When I visited the Lab in the company of Sally Coxe, on both occasions Kanzi jumped on the ledge near his keyboard and pressed SUE and BALL almost the minute we approached the glass partition. In my judgment it was purposeful—not an accident—and Jared's feeble attempt on the spot to explain it away I saw as pathetic or delusional." (LD)

10. A child can start or stop the movement at will. It appears to the child that it is free to control the voluntary actions of its limbs and bodies. Indeed, if the child were not controlling the movements of its limbs and body, who is? And why is the child driven to explore the relationship between the movements of its hands and feet, of the mobile hanging over its crib? Its hands and feet, controlled by its own "free will," in turn control the actions and movements of the mobile above it. Because if it does nothing and holds still, so does the mobile. Then, what prior conditioning and/or experience is determining the child to

desire to wave its hands and feet and watch them move the mobile? If we say that this behavior is "innate," we are saying nothing more than stating that the child does what it does because it is what it is. Or we could say it is what it is, because it does what it does. Either way, it is quite difficult to argue that the child is not moving its limbs of its own free will. If we wish to make such a claim, Determinism 1 does not work adequately, and we must turn to Determinism 2. (SSR)

11. See Sue Savage-Rumbaugh et al., "Ethical Methods of Investigation with *Pan/Homo* Bonobos and Chimpanzees," in Richard Gordon and Joseph Seckbach, eds., *Biocommunication: Sign-Mediated Interactions between Cells and Organisms* (London: World Scientific, 2017), 449–574.

12. Sue Savage-Rumbaugh, Kanzi Wamba, Panbanisha Wamba, and Nyota Wamba, "Welfare of Apes in Captive Environments: Comments on, and by, a Specific Group of Apes," *Journal of Applied Animal Welfare Science* 10:1 (2007): 7–19.

13. See Itai Roffman et al., "Preparation and Use of Varied Natural Tools for Extracting Foraging by Bonobos," *American Journal of Physical Anthropology* 158:1 (2015): 68–91; Itai Roffman et al., "Cultural and Physical Characteristics of Near-Arid Savanna Chimpanzees in Mali," *Human Evolution* 31:4 (2016): 191–214. In this latter article, Roffman et al. write: "*Pan* can realize their hominin potential both in a culture/language-rich environment and under extreme conditions in the wild."

14. Paul Celan, "Give the Word," in *Breathturn into Timestead: The Collected Later Poetry* (New York: Farrar, Straus & Giroux, 2014), 92–93.

Appendix

1. We also refer the reader to William Hillix and Duane Rumbaugh, *Animal Bodies, Human Minds: Ape, Dolphin and Parrot Language Skills* (New York: Springer Science, 2004). Despite its tendentious presentation of the work led after the 1960s, Gregory Radick's *The Simian Tongue: The Long Debate about Animal Language* (Chicago: University of Chicago Press, 2007) is a good source of information for the first part of the history of ape language research.

Index

(continued from page ii)

24 Without Offending Humans: A Critique of Animal Rights
Élisabeth de Fontenay

23 Vampyroteuthis Infernalis: A Treatise, with a Report by the Institut Scientifique de Recherche Paranaturaliste
Vilém Flusser and Louis Bec

22 Body Drift: Butler, Hayles, Haraway
Arthur Kroker

21 HumAnimal: Race, Law, Language
Kalpana Rahita Seshadri

20 Alien Phenomenology, or What It's Like to Be a Thing
Ian Bogost

19 CIFERAE: A Bestiary in Five Fingers
Tom Tyler

18 Improper Life: Technology and Biopolitics from Heidegger to Agamben
Timothy C. Campbell

17 Surface Encounters: Thinking with Animals and Art
Ron Broglio

16 Against Ecological Sovereignty: Ethics, Biopolitics, and Saving the Natural World
Mick Smith

15 Animal Stories: Narrating across Species Lines
Susan McHugh

14 Human Error: Species-Being and Media Machines
Dominic Pettman

13 Junkware
Thierry Bardini

12 A Foray into the Worlds of Animals and Humans, *with* A Theory of Meaning
Jakob von Uexküll

11 Insect Media: An Archaeology of Animals and Technology
Jussi Parikka

10 Cosmopolitics II
Isabelle Stengers

9 Cosmopolitics I
Isabelle Stengers

8 What Is Posthumanism?
Cary Wolfe

7 Political Affect: Connecting the Social and the Somatic
John Protevi

6 Animal Capital: Rendering Life in Biopolitical Times
Nicole Shukin

5 Dorsality: Thinking Back through Technology and Politics
David Wills

4 Bíos: Biopolitics and Philosophy
Roberto Esposito

3 When Species Meet
Donna J. Haraway

2 The Poetics of DNA
Judith Roof

1 The Parasite
Michel Serres

LAURENT DUBREUIL is professor of comparative literature, Romance studies, and cognitive science at Cornell University and distinguished professor of transcultural theory at Tsinghua University in Beijing, China. A former editor of *diacritics,* he has written ten books, including *The Intellective Space: Thinking beyond Cognition* (Minnesota, 2015).

SUE SAVAGE-RUMBAUGH is co-chair of Bonobo Hope, a nonprofit dedicated to increased understanding between apes and humans. She served as the lead scientist at the Language Research Center and later at the Iowa Primate Learning Sanctuary in Des Moines, Iowa, from 2004 until 2013. She is author or coauthor of more than 240 articles as well as many books and films.

Lightning Source UK Ltd.
Milton Keynes UK
UKHW012042280219
338204UK00003B/144/P